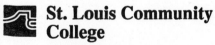

A Sense of PLACE,

a Sense of TIME

A Sense of PLACE, *a Sense of* TIME

John Brinckerhoff Jackson

YALE UNIVERSITY PRESS
NEW HAVEN AND LONDON

For Dorothée and Marc

"Pueblo Dwellings and Our Own" originally appeared as "Pueblo Architecture and Our Own" in *Landscape* 3, no. 2 (Winter 1953–54): 20–25. It was reprinted in John B. Jackson, *The Essential Landscape* (Albuquerque: University of New Mexico Press, 1985).

Designed by Rebecca Gibb
Set in Perpetua type by Tseng Information Systems, Inc.
Printed in the United States of America by The Maple-Vail Book Manufacturing Group
Binghamton, NY.

Library of Congress Cataloging-in-Publication Data

Jackson, John Brinckerhoff, 1909–
A sense of place, a sense of time / John Brinckerhoff Jackson.
p. cm.
Includes index.
ISBN 0-300-06002-5 (cloth)
0-300-06397-0 (pbk.)
1. New Mexico—Description and travel. 2. Landscape—New Mexico. 3. New Mexico—Social life and customs. I. Title.
F796.J27 1994
978.9—dc20 93-50618 CIP

A catalogue record for this book is available from the British Library.

The paper in this book meets the guidelines for permanence and durability of the Committee on Production Guidelines for Book Longevity of the Council on Library Resources.

10 9 8 7 6 5 4 3 2

CONTENTS

Preface

Most of the essays here offered were written over the last ten years, after I had ceased lecturing on landscape history. Several have appeared as chapters in books or in periodicals devoted to architecture or urbanism or landscape architecture: a wide variety of topics, but all related to my ongoing efforts to understand and define the contemporary American man-made landscape.

More and more of us share that interest: ecologists, planners, demographers, geographers, and economists; and the resulting literature dealing with the human use and abuse of the earth and its resources is impressive not only in its bulk but also in its urgency. Not all of it is of equal value. I myself have no liking for the cultural anarchy preached by the radical environmentalists. All too often their credo resembles that of some obscure and short-lived Christian heresy in which the cross would be interpreted as the symbol of the dismembered forest tree, and death is seen as merely the first step in the producing of compost for the man-ravished planet. But insofar as environmentalism betrays an awareness of a cosmic order, it can be welcomed as one more sign of a renewed religious interpretation of our actions as inhabitants of the earth.

The study of landscape history contributes its share to the new approach by reminding us, among other things, that since the beginning of history humanity has modified and scarred the environment to convey some message, and that for our own peace of mind we should learn to differentiate among those wounds inflicted by greed and destructive fury, those which serve to keep us alive, and those which are inspired by a love of order and beauty, in obedience to some divine law. The whole inhabited world, from the highlands of Peru to the heart of Asia, is marked by vast circles and parallel lines and spirals, great avenues of monoliths, many dating back thousands of years: signs of our sense of responsibility for the survival of the earth and its people. Could much the same not be said of our immense grid landscapes, our geometrically designed cities, our parks and wilderness areas?

The question is not always easy to answer, and landscape students have no special insight. Sometimes the desecration of a landscape is so ancient that we accept it as part of nature; whereas the most conscientiously designed improvements of recent years can be rejected indignantly by purists. All serve as symbols; symbols of what we have done that is wrong or of what is appropriate and right. My own search for the more popular, more everyday symbols in the American landscape has often led me far afield: into Pueblo Indian villages, rural trailer communities, into parts of the city dominated by industrial traffic, and into the railroad-oriented towns of the High Plains. I made no great discoveries and wound up in plenty of blind alleys, but as my text indicates, I have come back convinced beyond a doubt that much of our contemporary American landscape can no longer be seen as a composition of well-defined individual spaces—farms, counties, states, territories, and ecological regions—but as the zones of influence and control of roads, streets, highways: arteries which dominate and nourish and hold a landscape together and provide it with instant accessibility. This means, I think, that architecture no longer provides the important symbols. Architecture in its oldest and most formal sense has ceased, at least in our newest landscapes, to symbolize hierarchy and permanence and sacredness and collective identity; and so far the road or highway has not taken over those roles. The road generates its own patterns of movement and settlement and work without so far producing its own kind of landscape beauty or its own sense of place. That is why it can be said that a landscape tradition a thousand years old in our Western world is yielding to a fluid organization of space that we as yet do not entirely understand, nor know how to assimilate as a symbol of what is desirable and worth preserving.

In time, we will find our way and rediscover the role of architecture and man-made forms in creating a new civilized landscape. It is essentially a question of rediscovering symbols and believing in them once again. Many centuries ago, when Britain was no longer Roman and not yet Christian, a nameless poet was one of those who glimpsed the symbolism of the surrounding scattered remains:

> *Wonderful is this wall of stone,*
> *wrecked by fate.*
> *The city buildings crumble,*
> *the bold works of the giants decay.*
> *Roofs have caved in, towers collapsed,*
> *Barred gates have gone,*
> *gateways have gaping mouths,*
> *Hoarfrost clings to the mortar.*[1]

Out of a ruin a new symbol emerges, and a landscape finds form and comes alive.

1 The Accessible Landscape

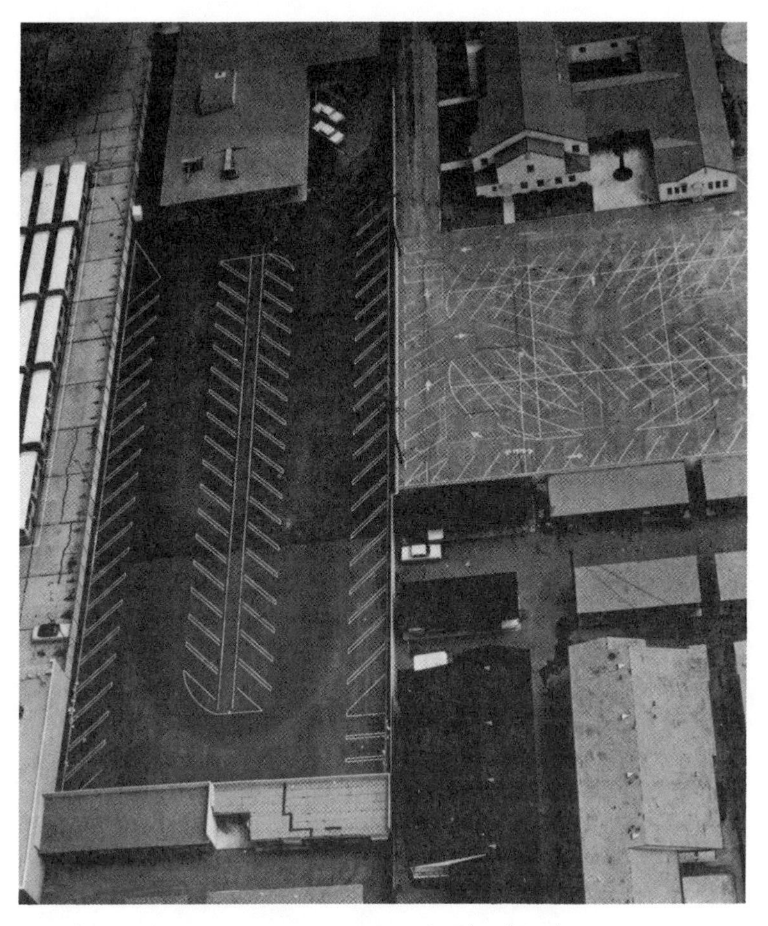

Parking lots from *Thirty-four Parking Lots in Los Angeles* by Edward Ruscha, 1967.

Those of us who are old enough to recall the world of seventy years ago can find some satisfaction in having played a role—minor and entirely unheroic—in the current exploration of space: we were the first generation to become accustomed to seeing the earth from the air.

It was as passengers that we flew, to be sure. Hundreds, perhaps thousands of genuinely venturous flyers had already confronted new and unknown dangers and had acquired the hard way new skills, new insights, and a new self-awareness. It was they who first experienced the wonders and terrors of the ocean sky and were not deterred. All that we did was gaze down on the earth beneath us and idly speculate about our new relationship. The planes we flew in were small, seldom carrying more than eight, and we seldom flew very far. No matter; we were pioneers of a sort, and there were very few of us. Even as late as 1926, not more than eight thousand people a year were bold enough to go out to a grass-grown landing field and entrust their lives to a one-propeller plane flown by a young pilot wearing goggles.

The takeoff was an event I will never forget, but the real excitement came when I found myself looking down at some part of the landscape of the Midwest or Great Plains. Flying in the East meant seeing familiar landmarks from an unfamiliar perspective; it was only when I looked at the multicolored pattern of rectangular fields and checkerboard towns, repeating itself over and over from one horizon to the other, that I discovered the typical American landscape.

In the early days of commercial aviation many first-time flyers were puzzled by the prevailing rectilinearity and assumed that it was the work of unscrupulous real estate operators laying out some monstrous megacity. I might well have thought the same, only I remembered something I had learned in school about the National Land Survey of 1787, and how Jefferson and others had devised a plan whereby all the vacant unclaimed land in the young republic could be divided into an almost infinite number of squares, each of them a square mile, or 640 acres—

The grid divides farm land in the Red River Valley, North Dakota. (Photo © Alex MacLean)

more than enough to satisfy the average would-be settler. The so-called grid system, oriented to the four points of the compass and extending from the Ohio to the Pacific, could also be organized into townships and counties and even states, all beautifully rectangular. It was an ingenious, if unimaginative, way of creating a landscape, but it was not easy to get used to, and in the beginning settlers from the East or from Europe complained of its monotony and its disregard of the topography; in fact, the grid made no adjustment to rivers or hills or marshlands. Still, by now, more than two centuries later, there are millions of Americans so thoroughly at home in the grid that they cannot conceive of any other way of organizing space. I have been in homes in Kansas where they refer to the southwest (or northwest) burner on their stove. They tell you that the bathroom is upstairs, straight ahead south.

We have to realize that the grid system was never meant to be a way of laying out cities and regions, never meant to produce close-knit communities: those were for the settlers to create. Its purpose was merely to facilitate the distribution of land as simply and as equitably as possible; it was the most direct way we had to see that farmers could acquire land for a farm—even if it meant much loneliness before the first neighbors appeared on the scene. For to have a clear title to a piece of fertile land was at that time a dream of every American. Many changes in the composition of those square townships have taken place since then, but whenever I looked down at the vast array of rectangles and squares and stripes, at the scattering of white houses and red barns and grain elevators, at the villages in the midst of trees, I took some satisfaction in realizing that for many families over the generations the dream had become reality.

I am sure that every other passenger was impressed by the spectacle and felt something like patriotic pride. In those days we all still retained a somewhat simplified image of traditional rural America: a composition of green or pale-yellow or dark-brown fields; of clusters of houses surrounding a white steeple; of placid streams and a glimpse of wooded hills. We had inherited a veneration of private land, of private territories, each of them distinct, bordered by fences or rows of trees; a landscape remote from factories and cities. Once reassured that the grid system was *not* a real estate scheme, we saw it as reassuringly permanent in its attachment to land and family.

Roads, of course, were part of that landscape, at least in theory, but they were not always easy to discern. In that original survey no provision had been made for the building or even the establishing of roads. In the pioneering years wheeled vehicles were scarce, and the first settlers to arrive either made use of what Indian trails there were, or came by boat or raft. Once settled on their piece of land, they had to find their way—either on foot or on horseback—to the nearest village or outpost if they wanted company.

In time a grid of roads or trails evolved along the lines determined by the survey, but for many years these roads were rough, and often deep in mud. It was commonly said in country newspapers that roads were essential for citizens who wanted to attend lectures, go to church, or pay taxes, but a grid system of roads rarely led directly to the village or county seat; the roads were lonely and neglected, and rarely used. Even today in the Great Plains, where landholdings often comprise several townships, there are wide areas of undulating prairie entirely without roads of any sort; the arcadian landscape is marked only by grazing cattle as they move toward a watering place.

One of the least investigated aspects of our European-American culture is our ambivalent attitude toward the road and the street. In their infrequent mention of roads, historians and even many geographers tend to adopt the establishment point of view that roads are essentially for the maintenance of order and for commerce (or warfare) with neighboring states. Nevertheless, there has always been and probably always will be a widespread distrust among average men and women of all roads which come from the outside world, bringing strangers and strange ideas. Reactions to such roads vary from age to age, from one region, one class, one stage of economic or social development to another; yet underlying all those variations, there seems to be a basic human response: the road is a very powerful space; and unless it is handled very carefully and constantly watched, it can undermine and destroy the existing order.

I have a theory that about four or five centuries ago, during the Renaissance, the Western world learned how to organize the spaces we lived and worked in so that they would achieve what appeared to be the most efficient, the most enduring, the most beautiful design: spaces in the house, in the garden, on the farm, in towns and cities, even spaces which were whole nations or empires. To assure the survival of that spatial system, one thing was essential: we had to discipline those meandering, unpredictable roads and paths and alleys and trails which had proliferated since the beginning of history and which, like a web of roots, always threatened to heave up and ruin even the most carefully planned landscape of spaces. For it was generally agreed that the only useful roads were those which the establishment had built to serve military or commercial purposes.

It would be quite possible to write a landscape history that would describe our various efforts to discipline roads and streets, but it would be a history of countless setbacks and new beginnings, making for discouraging reading. For whenever the establishment begins to congratulate itself on having subdued that anarchic instinct, some unforeseen development or event emerges to give the road a new explosive vitality, and the landscape is threatened with confusion.

The seventeenth century (to go no further back in the chronicle) saw itself as the climax of civilization; then there came a rapid increase in the number of wagons and carriages meant for relatively long-distance travel on important errands. Almost at once a new and expensive kind of road was called for to accommodate wheeled vehicles. At another time new types of farming to feed the urban population evolved: roads and trails for the movement of produce and animals had to be created. Civil unrest necessitated the building of a whole network of reliable roads in prerevolutionary France, and when the eighteenth century discovered the picturesque beauties of the rural landscape, smooth, well-drained roads to accommodate the light, open carriages became a priority. What the advent of the automobile, now a hundred years in the past, has done and continues to do to our old system of roads and streets is a familiar story. It is enough to say that our cities and our countrysides have only begun to adjust to the new situation.

Why have we not foreseen these changes? Why have we not wholeheartedly welcomed each new kind of road as evidence of progress? Because two traditions persist to this day: the Renaissance assumption that the system of spaces has priority over the road as a tool of change, and second, a deep-seated belief among the mass of men and women that roads are meant to serve the small community, not the state. We still perceive the landscape (especially from the air) as designed to suit the tastes and needs of an element in society that geographers call *homo dormens*—man the owner or occupant of land, man the territorial animal who will defend his place to the end, and who erects fences and walls and frontiers to keep out the intruder and maintain landscape stability. In such a view of the world, roads are of secondary importance. On his first trip through New England as president, Washington complained that the roads twisted and turned in order to follow property lines, and were impossible to follow. On that same trip, Washington was approached by the beedle (an obscure church officer) in a small village and notified that he was not to travel on Sunday on the village roads, such being the local law to preserve the sanctity of the Sabbath. Washington responded tactfully by saying that his horses needed a day of rest. But the episode reminds us of a long-forgotten fact: that the archetypal road is one which not only serves the daily needs of the small community but helps preserve its ethical values. It is essential to the work routine and the routine of worship and celebration, but more than that, it makes virtuous behavior possible and it preserves the territorial integrity of the village. Van Gennep cites numerous instances of villages where certain rites are observed when a traveler crosses the village boundary as he approaches by road.[1] The road as an important part of the village calls for such a rite. The territory is not to be entered without a form of initiation.

The modern world has largely done away with such local roads, but we do not have to look very far to discover that territorial instinct in

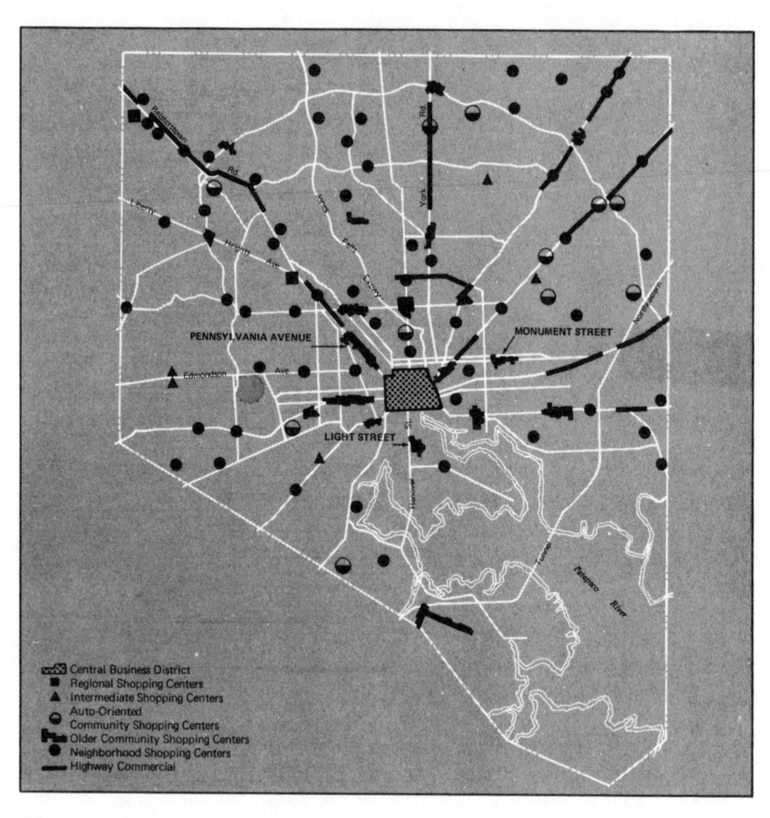

The strip and its inner-city successors: auto-oriented centers in Baltimore. (From Grady Clay, *Close-up,* Chicago, 1973)

regard to roads is still alive. City gangs define and mark their territory by means of graffiti on strategic street corners and walls. Recently in New Mexico a small community of working-class mobile homes, indignant at the speed of passing trucks and the dust they raised, organized and erected a large sign declaring that the legal speed limit on "their" road was eighteen miles per hour, and threatening penalties. The police soon put an end to the action.

Sooner or later in the modern landscape every country road leads to a highway, and all highways, whether we follow them or not, lead to the city. All flights, however distant and adventurous, eventually end at some municipal airport where we never fail to be overwhelmed by the scale and immensity of the new spaces that have emerged over the last generation: the endless runways, the enormous parking lots, and the view of the cluster of highrise buildings in the downtown city. It is here in the city, not in the open countryside, that that modern road reveals itself. Even a generation ago the old architectural order prevailed: the street was still squeezed between tall and imposing facades—the urban equivalent of those fences flanking the country road: protecting what lay beyond. But the new road or street is like the eight-hundred-pound gorilla: it goes where it wants to. It is wider than roads were in the past, sinuous in its layout, no longer respectful of the grid, and it devours spaces and structures hitherto thought of as sacred. It is creating its own architecture: short-lived, eager to conform to the new type of traffic and to discard its old symbols and any hint of history.

The automobile of course is what has determined the dimensions and capacity of the new street, and architectural critics and Americanists of all descriptions have been quick to analyze its impact on the life and appearance of the city: a whole library of books and films has been produced telling of how and when and where the new street has replaced the old. But the real consequence of the rebuilding of our cities is not architectural: it is the emphasis on accessibility, the gradual but total destruction of the distinction between the life of the street and the life lived behind the facade. What has taken place is the elimination of those immemorial rites of passage that were once the hallmarks of our culture. Those architectural monuments—the church, the university, the office or place of work, the public building, the restricted residential area, all once characterized by a degree of isolation and internal privacy, are now wide open and accessible to the street.

The elimination of rites of passage started modestly enough with the drive-in business. Then the territory became a parking lot; the church became a building open to all, offering instant salvation; the library let us into the stacks; the public building welcomed us with consumer-friendly decorations; the supermarket did away with clerks; the hospital let us wander at will. The doorman, the receptionist, the head waiter, the host all vanished, and the only rite of passage that survived was the insertion of a magnetized credit card. But these are the more superficial aspects of the new accessibility, and they have been sufficiently commented on. What we have still to realize is how the new aggressively vigorous street life has produced new and much simplified work procedures and new

The private domain. (Photo: Barrie B. Greenbie)

mobile types of work and production, and how the inner-city has been eviscerated by the accessibility and the destruction of protective rites of passage. And we have yet to do more than lament the transformation of the countryside by the superhighway.

The question which insists on an answer is, What kind of small or local community can we hope to have? What we can be sure of is that it will not be based on territoriality. What seems to bring us together in the new landscape is not the sharing of space in the traditional sense but a kind of sodality based on shared uses of the street or road, and on shared routines. The trailer court does not last for many years and its amenities are scanty: but it is accessible, and the highway is accessible, and if somewhere jobs are accessible, we can look forward to something like a community held together and achieving identity by the short private road which everyone uses in daily life.

SOUTHWEST

2 Seeing New Mexico

William Clift, La Mesita from Cerro Seguro, New Mexico, 1978. (© William Clift 1978)

We learn about history by reading it in school; we learn to see it when we travel, and for Americans the place where we see most clearly the impact of time on a landscape is New Mexico.

Our New Mexico history is more complicated than most, and far more visible. In regions more prosperous and blessed with more abundant rainfall, the past, even the recent past, soon vanishes from sight: bulldozed out of existence in favor of something new and more costly, it is also often quickly hidden by exuberant vegetation. Even the rubble of abandoned tenements in the Bronx soon acquires a covering of weeds and vines and wildflowers, and trees conceal the abandoned farms of Appalachia. But in New Mexico history remains exposed for all to see. Our landscape is everywhere spotted with ruins—ruins of ancient towns, ruins of sheepherders' shelters built a decade ago. It is as if we had been struck by a neutron bomb, eliminating people while leaving their dwellings intact, at the mercy of wind and sun. It is to see our past that thousands of tourists come to New Mexico: archaeologists, geologists, antiquarians, lovers of whatever is old or out-of-date or mysterious because of age. Our history invites the photographer.

The best time for seeing history is the summer. That is when the remoter country can be explored; it is when schools and colleges all over America are closed, and teachers and students and scholars are free to wander. It follows, therefore, that awareness of southwestern history is a seasonal phenomenon determined by the academic calendar, much as a certain kind of piety is determined by phases of the moon. Summer is the time for looking back and recording what we see. Family reunions, two-hundred strong, gather in the shade of a cottonwood grove, in a dance hall, in a half-forgotten village once an ancestral stronghold. Veterans' organizations parade down Main Street, and Santa Fe and Wagon Mound and Arroyo Seco deck themselves out in Indian or Spanish-American or cowboy or counterculture costume, celebrating the Old Days. The sun shines out of a deep-blue sky; it is hot, but not too hot, and history is transformed into a photographic pageant, ideal for color slides.

Yet it is hard not to be fleetingly aware of a background suggestive of a kind of history with a different dimension, no matter where we are in New Mexico. We glimpse it in a dark face in the crowd; we catch an echo of it in the voices and the music coming from a corner bar. We see and never quite forget the horizon of range after range of mountains of diminishing blue. In every background there lurks another kind of past, far less easy to comprehend than the strictly human history on display.

There is one region of New Mexico where we can come close to time measured not by events or seasons but by millennia, a landscape with a history that is perhaps not history at all, merely the unending repetition of cosmic cycles, a landscape where by a paradox the still photograph records all we can ever know of its past. The Colorado Plateau is the name given by geographers and geologists to an immense region covering most of Utah, western Colorado, eastern Arizona, and northwestern New Mexico. When you drive due west of Albuquerque toward Grants and the uranium country, you catch a glimpse of a small portion of it—a horizon like a long, pale-blue rampart, extending beyond sight to the north and south. It is deceptively unspectacular, almost a continuation, one would say, of the pleasantly humanized landscape of the Rio Grande country. But this is actually only the eastern edge of a province distinguished by its great elevation (reaching in places to eleven thousand feet), its hundreds of remarkable canyons (including the Grand Canyon), and its overall horizontality. Every mesa, every canyon, every freestanding peak seems composed of layer upon layer of red and brown and yellow and dead-white rock.

Only a small fragment of this imposing landscape lies in New Mexico, but it is a fragment containing some of the largest prehistoric ruins in the United States, as well as areas with occasional stands of trees and small streams meandering through canyons. There are expanses of pale grass and sagebrush, and piñon and juniper trees in groves on the slopes of the valleys. It seems to be empty of life, but in summer it sometimes has a pastoral, almost arcadian quality. Navajos graze their flocks of goats and sheep on the grass among the miniature forests of sagebrush, bells tinkling. In the middle of the day they rest in the dense black shade of the piñons. The air is fragrant, the light on the perpendicular, dark-red canyon walls is golden. Small clusters of ragged Navajo dwellings, with a peach tree or two nearby, stand under the piñon trees, and a saddle horse sleepily hangs his head. Turn elsewhere and the view is perhaps a little too vast for comfort: a panorama of endless range country with a rim of violet mesas and dark mountains where there must be forests and streams of water, though far away. The days are all alike; the summer is long and immobile. In the late afternoon immense black clouds boil up to the zenith, and then some small portion of the hot and thirsty landscape is suddenly blessed with a brief, violent downpour which makes every rock, every patch of earth glisten. The storm comes to an abrupt end like a duty routinely performed, and is followed by not one but two perfect rainbows. It is as if some rite had been reenacted, some myth made visible for the millionth time, antiphony to a ceremonial dance in a nearby Indian village.

Which comes first: the blessing or the prayer? It is not easy in this landscape to separate the role of man from the role of nature. The plateau country has been lived in for centuries, but the human presence is disguised even from the camera's eye. There are ruins like geological formations, disorders of tumbled stone. There are immense arrays of slowly crumbling rocks that look like ruins. The nomenclature we white Americans have imposed on much of the landscape testifies to our uncertainty: the ruins have unpronounceable Navajo names; the natural formations are called Gothic Mesa or Monument Valley or Chimney Rock.

It is the sort of landscape which (before the creation of the bomb) we associated with the world after history had come to an end: sheep grazing among long-abandoned ruins, the lesson of Ozymandias driven home by enormous red arches leading nowhere, lofty red obelisks or needles commemorating events no one ever heard of, symbols of the vanity of human endeavor waiting to be photographed. But is that really the message of the plateau country? There was a time toward the end of the last century when photographers, masters of their art, had a more precise vision: they wanted to leave history, even human beings, out of their pictures. Perhaps there were technical reasons for wishing to exclude all movement; perhaps it was a matter of belief, a way of responding to the concept of time in the Colorado Plateau. For what makes the landscape so impressive and beautiful is that it teaches no copybook moral, no ecological or social lesson. It simply tells us that there is another way of measuring time and that the present is, in fact, an enormous interval in which even the newest of man-made structures are contemporary with the primeval. That is why it is possible to see that the dams on the Colorado and San Juan rivers, the deep pits of the Santa Rita copper mines, and the terraced mountains near Laguna where uranium has been extracted are all as old or as young as the canyons and mesas and the undulating plains of sagebrush.

Not far from Quemado (which is not far from the Arizona line) there is a field of innumerable lightning rods, geometrically planted in an expanse of range grass. As an example of contemporary environmental art it is a source of infinite curiosity and bewilderment. Some day, centuries hence, the field of lightning rods will have been forgotten by tourists and entirely assimilated into the landscape. Navajos grazing their sheep among them will know that these rods derive from the same cosmic occurrence that balanced liver-colored rocks on pedestals of yellow mud in the Chaco region: objects identified with an Emergence myth, easily explained, provided our small-scale microhistory is left out of the picture.

That school of "timeless" photography flourished at a period when all of New Mexico was described by outsiders, even admiringly, in terms of its peculiar notions of time. It was "the land of poco-tiempo," "the land of mañana," "the land where time stood still." What was meant was not Indian or prehistoric New Mexico, but Spanish-American New Mexico.

By and large this is the New Mexico associated with the upper Rio Grande Valley and the mountains containing it. It was here that the first colonists settled in the late sixteenth century, and it was here that the province (or state) acquired its identity. What attracted settlement was the mild climate, the apparent abundance of water, the fertile soil, and the forests covering the mountains. In many ways the landscape seemed to resemble that of Spain. Almost from the time of the first explorations New Mexico was seen as a kind of promised land: not a paradise of ease and abundance, to be sure, but a land of grass and forests and flowing water where the efforts of working men and women would be duly rewarded. For it so happens, even today, that no matter whether you come to New Mexico from the immediate east, the High Plains, the arid south, or the canyon landscape in the west, the region always seems, by comparison with the country you have been traveling through, a land flowing with milk and honey. What shatters the illusion is the long dry summer that afflicts the greater part of the state.

How long it took the earlier generations of Spanish-speaking colonists to learn that lesson is a complicated question: the presence of hostile Indians in the plains of the eastern part of New Mexico discouraged their settlement and even exploration until the mid-eighteenth century. In any event, Spanish settlement was long confined to the Rio Grande region, which to this day remains the heartland of Spanish-American culture. The small lateral valleys of the river, as well as the valley of the Rio Grande itself, provided the colonists with an environment suited to their kind of agriculture and their kind of living—in small villages where old-established customs and relationships could be continued. Settlement in colonial New Mexico was in effect a transplantation, a new version of the order that had prevailed in colonial Mexico and Spain. It was not the work of footloose individuals in search of adventures or wealth, but of small, homogeneous groups of simple people who brought with them their religion, their family ties, their ways of building and working and farming.

Farming meant irrigation; to that extent the settlers were aware of the climatic limitations of the region, and knew that the only places where that kind of farming was possible were along the few permanent watercourses in the foothills and valleys. Each village devised its own communal irrigation system—an accomplishment deserving of more recognition than it has so far received; and each village created its own miniature landscape of gardens and orchards and fields and pastures, a landscape distinct from the surrounding wilderness. Farmers not only introduced new kinds of vegetation—crops and grasses and fruit trees—but also another climate, for their irrigation systems made them relatively independent of the unpredictable local rains.

The history of these villages is largely unrecorded; all we usually know about them is roughly the decade of their settlement, the date of the first church, and the place of origin of their first settlers. Indian raids, feuds with neighboring villages, the building of a road to the outside world—important events in their time—remain a matter of legend or hearsay. The destruction of the irrigation system by a cloudburst, the erosion of fields, the incremental destruction of the nearby forests, and the desertion of the village itself—these are confirmed by visual evidence. But what is lacking is any picture of the villages in their prime. Those of us who are old enough can remember places in the foothills of the Sangre de Cristos or in the valleys of the Rio Puerco, the Pecos, the Rio Grande as they were a half-century ago. They had already begun to decline, and signs of increasing poverty and depopulation were painfully clear, yet there were still cultivated fields and well-kept irrigation ditches; there was a general store, there was a school, there was a freshly painted church and a neat graveyard. On Sunday afternoons the young men of the village and from the nearby ranches, dressed in finery, galloped up and down the only street. There were still men and women who could identify the village a stranger came from by his or her accent, who knew the local name for every field, every hill, every wild plant. They knew their landscape by heart.

One after another, over the decades, the settlements died, but not without resistance. A flood buried gardens and fields under gravel or sand; a local resource—wood or game or a special crop—lost its market; a railroad ceased operation; the school was closed. Rather than abandon their home, the villagers became ranchers and raised cattle or sheep. But in the end it died, and others died: first the remote villages on the margin of the plains, where there were no other jobs, and then the villages where the rangeland had deteriorated and the cedars and junipers were coming back into the abandoned fields. All that is now left of that traditional farming landscape are the villages in the mountain heartland and in the Rio Grande Valley.

Time in those secluded places has a special flavor—a resigned, slow, autumnal rhythm. The colors linger into the early winter, in the brown and orange leaves on the cottonwoods along the streams and irrigation ditches, in the strings of red chili on the fronts of houses, and in the groves of lemon-yellow aspens far up in the mountains. Then a winter wind sends all their leaves to the ground in a shower of gold, and the chamiso turns grey.

Snow that lasts comes in later November and remains on the higher slopes of the Sangre de Cristos and the Jemez until well into the spring. In the valley and foothills it slowly melts, leaving patches hiding under the piñon trees, but in the heights and in places the sun reaches only for a few hours a day, winter is a season to be taken seriously. It transforms the smaller dirt roads into lanes of bottomless mud. The rancher stays close to headquarters, and villagers think twice before driving their mud-splattered pickups into the forest after firewood; even in town we are careful to stay on the paved surfaces. What was recently a landscape of coming and going and outdoor work—a landscape of gardens and orchards and small farms—almost overnight has turned into a scattering of isolated villages and hamlets. The cold and the wretched roads make every community, every family shrink into itself, and the silence is rarely broken. In the old days the clanking of tire chains was

Anonymous photograph c. 1930 of Cordova, New Mexico. (Courtesy of the Museum of New Mexico, Santa Fe)

part of winter in the country, but in the mountains of northern New Mexico, as elsewhere, we no longer hear it, and the almost perfect soundlessness is what visitors notice first of all. Find out for yourself what this means: stand on a hillside overlooking a village of tin-roofed houses on the edge of the forest in the Sangre de Cristos or in some part of the Pecos Valley; if it is a bright day in January or February you will hear the screaming of flickers in the groves of piñon. Then in a backyard, perhaps a half-mile away, someone is chopping wood. Go down into the village where there is the familiar sound of snow melting off the tin roofs. Not a voice is heard; life has withdrawn into the houses behind closed doors, and the windows, with geraniums in tin cans, are obscured by frost. Someone tries to start a car but soon gives up. In the cold, starry night the lights are few and dim, and you can barely make out the landscape of black forest and small, snow-covered fields. If you are lucky you may hear, late at night, the yelp of a coyote. It sets the village dogs into a frenzy of barking.

It is hard to remember, despite all we have read about the history of this landscape, that as the crow flies (or as the car travels) Mexico, once the motherland, is not distant. But it is separated from us by more than barriers of mountains and desert: climate, once summer is past, abruptly inserts a northern winter which produces two contrasting worlds, two contrasting ways of living that last until well into spring.

One of the happiest moments in our century was when we discovered the landscape of ice and snow, and made it into the setting of sports and cross-country adventure. This may well have originated here in North America; a French geographer, Pierre Deffontaines, traveling for the first time several decades ago in wintertime Canada, noted to his delight how the snow-packed roads and the frozen rivers fostered a constant coming and going by sleigh from one village to another and a whole season of sociability and outdoor activity. In summer farm work, slow travel by wagon discouraged hospitality; enforced winter leisure meant family reunions, excursions, races, skating and hockey and skiing, and fishing through the ice. Urban holiday makers in Europe and America soon followed suit, but there were always areas of rural resistance: Ethan Frome in rural New England and the villagers of New Mexico remained loyal to winter as a season of environmental Calvinism, a gloomy variation in the Latin religious calendar that briefly cuts the historic ties with that gregarious society south of the Rio Grande; and not even the renewal of outdoor public life in summer can entirely dispel the silence of winter.

Climate, no less than an ingrained sense of what is fitting, clears the plazas and the lanes of the last summer idlers, one leg propped against the wall, talking with grave voices. Climate, coupled with loyalty to family, keeps us home where we sit in silence, pondering old grievances and searching our souls. Outside, the clear bright air smells of snow and piñon smoke; inside it smells of coffee and roasting chili and wet clothes drying near the stove. Climate, sooner or later, makes us return to origins, makes the tourists and environmentalists and students of folklore and handicrafts scurry back to Berkeley or New York or Dallas to show their brightly colored slides of the Land of Enchantment and dream of owning an adobe house of their own, with hollyhocks in the front yard, and a loom or a potter's wheel or dulcimer in the cool, dark room within. Climate tells us to stay where we belong and to do what we have always done. On Sunday (in remoter, smaller villages every other Sunday) the cracked church bell sounds off with an unmelodious bang! bang! bang! A stove in the corner crackles and shines but fails to heat. After the service there are brief greetings on the church doorstep, yet nothing in winter can keep us together for long. That is the virtue and even the beauty of this time of year in northern New Mexico: it isolates and intensifies existence; it creates a landscape and then preserves it by freezing it.

nineteenth-century explorers who believed that the desert began somewhere in eastern Kansas. To them any region without trees and not adapted to traditional eastern methods of farming was desert. Much of New Mexico, in fact, can be called arid or semiarid—an immense, rolling, underpopulated country covered by short, wiry grass, which in the early summer turns the color of straw.

Geologically speaking, much of the eastern half of the state is an extension of the High Plains—of the Texas Panhandle and Oklahoma. But what distinguishes it from that impressively monotonous region is its variety of landforms—innumerable, widely scattered, dark, steep-sided mesas, floating on the sea of pale-yellow grass like a fleet of flat-tops riding at anchor; the cones of extinct volcanoes; the many canyons. These last are remote and hidden from view, and those who formerly explored the rangeland on horseback rather than from the air came upon them with frightening suddenness. All that betrays their existence is a scanty fring of piñon and juniper on their rim. You find yourself gazing down into a long, deep, narrow valley with almost vertical walls of red or brown rock, and below where you halt there are the tops of cottonwood growing along some meandering stream for hundreds of feet.

These enormous landforms are about the only variety the landscape of New Mexico provides. In the spring, long after the winter snows have melted and left pools of clear water in the hollows of the rangeland, the grass is a brilliant green, and the expanses of wildflowers—there are said to be more than six thousand varieties in the state—are spectacular. But much depends on when you see the eastern region. If in April, it seems to be potentially ideal farming country; in July it is a sun-baked emptiness, to be avoided whenever possible. Those of us who live here the year round are well aware of the seasonal change. Our lives revolve around the man-made elements in the landscape. We shuttle between people and places—specific people and the specific places where they live and work and relax. The expressionless solitudes of the open road between, let us say, Vaughn and Roswell, or Tucumcari and Hobbs or Logan, occupy—or used to occupy—little of our thought. We learned to welcome almost every trace, every sign, no matter how incongruous or unsightly, that reminded us of the human presence: the lonely two-pump gas station, the gate and cattle-guard entrance to some far-off, invisible ranch, the tattered billboard out in the middle of nowhere. We were (and perhaps still are) attracted to ruins, no matter what their size

Decay is the negative image of history, and its presence througl
ern New Mexico has long fascinated the wandering photograpl
the essence of Spanish-American rural culture. The relentle
of ruin and abandonment has been interpreted as a kind c
flowering, something to be recorded before it is too late. Th
fact, a period after World War II when the landscape of the
Cristo villages and the upper Rio Grande Valley was seen ex
a panorama of crumbling adobe walls, sagging roofs, doorw
doors, abandoned roads bordered by rusty barbed wire, lea
tainly to overgrown fields and resurgent forest. There was r
except the old and defeated, never a sign of continuing life
sad pictures of deserted graveyards. This vision, repeated l
many other parts of the country, seems in retrospect to have
reflection of reality than a way of expressing a nostalgic ver
tory: a desperate, last-minute recording of old and once cheri
the New Mexico chapter in that once popular chronicling o
America," the old America of small farms and villages and s
fields. We captured on film the ghosts of places not yet enti

As long as those remnants of nineteenth-century N
survive as more or less recognizable human artifacts, they wi
of the old order—and of an older photographic approach
world. But it is increasingly evident that young Americans a
to discover the new landscape that is evolving, demandin
tion and interpretation, if not necessarily our critical accepta
has started a new chapter, and our vision expands to incluc
landscape.

Actually, it is not a new landscape; it is an aspect
tial New Mexico landscape—hitherto empty and forbiddir
been explored, invaded, and occupied. In the last generation
the first time, ventured out beyond the familiar, protective
watered valleys and forested mountains, beyond the green
rain and snow and the traditional succession of seasons, and
taken the settlement of the semiarid plains, the naked mc
the deserts of the Southwest.

Desert is not a word people in New Mexico like to h
used. It hurts us to read in eastern papers references to
around Santa Fe or to the "desert" climate of Albuquerqu
is intended. The term conjures up a pleasant image of sile
tery and beauty, and its use is a carry-over from the wri

or age. Their shabbiness served to bring something like a time scale to a landscape, which for all its solemn beauty failed to register the passage of time.

The story of the dying and small rural communities in every part of the world has become familiar to us over the last century and a half. It is most impressive, most regrettable when it tells of the decay of a well-known and well-loved landscape, like that of New England or New Mexico, but the moral of the story is in almost every case the same: existence for people in the country became more difficult, more joyless and without reward. Low pay, monotonous work, a sense of being isolated and forgotten, of diminishing hope for the future afflicted one village, one farmstead after another. For more than a century we have seen it happening, so it is not too early for us to look elsewhere in the countryside and to become aware of the new communities, new installations that are evolving in every rural landscape. If much of the migration from that landscape in New Mexico has found its way to large cities, much of it, perhaps most of it, has swelled the population of small towns and even created entirely new types of settlements—still rural in location but essentially industrial or commercial in economy, dependent not on a stream or river or a climate of familiar seasons, but on a highway, a dam, a mine, a tourist attraction. The movement away from the countryside is everywhere, but in the relatively empty landscape of New Mexico the fluidity is more easily discerned, and we can see more than the decline and death of the traditional order. We can see the emergence of a new kind of community—new in that it represents a different relationship with the environment, a deliberate confrontation with elements in the landscape that earlier generations sought to avoid.

The same invasion of the arid or desert environment is taking place in Arizona and California. Wherever we can, we seek out a fresh, untried, unknown setting and impose a new technology upon it, a new grasp of environmental factors. Local attachments, roots, are not what we are after: resources will probably be exhausted, tastes will change, and better jobs will lure us to another part of the desert or the mountains, or even back to the city. What New Mexico seems to offer is what it has always offered: the dramatic confrontation between the new and mobile and optimistic human installation on the one hand, and the overpowering "timelessness" of an ancient landscape with its cosmic chronology on the other.

3 Pueblo Dwellings and Our Own

Mesa Verde cliff dwellings. (Photo: Barrie B. Greenbie)

The territory occupied by the prehistoric Pueblo culture can briefly be described as the immense, scantily populated plateau country centering on the Four Corners where New Mexico, Arizona, Utah, and Colorado touch. It is uniformly high—the elevations of the Pueblo villages range from five thousand to eight thousand feet—and its rainfall is seldom more than fifteen inches a year. The Hopi villages of Arizona survive on less than eleven inches. By contrast, the rainfall of eastern New England is about forty-three inches.

There are few perennial streams, though small springs are numerous, and the landscape is a vast and colorful panorama of mesas and canyons and broad valleys. Except in the few spots favored with water, the vegetation is harsh and thin, but no village is more than forty or fifty miles from mountains with stands of pine and spruce and aspen, and an abundance of game. The dry and sunny climate, punctuated by brief, violent storms, is remarkably healthy.

For all its common environmental characteristics, the area has enormous diversity. It includes the hot lower Rio Grande Valley, where forests are all but unknown and the growing season is long, as well as the cool foothills of the Sangre de Cristos. There are half-hidden valleys with fertile soil and flowing rivers, and in the western part wide plains choked by rock and sand. In any Old World region of similar extent we would see a great regional variety of cultural landscapes, no two of them alike, but in the prehistoric Southwest cultural variations are surprisingly few, and variations in architecture even fewer. "The essential uniformity of [building] types which prevails over the immense area covered by the ancient Pueblo ruins," Cosmos Mindeleff remarks, "is a noteworthy feature, and any system of classification which does not take it into account must be considered as only tentative."[1]

Since these words were written, almost a century ago, archaeologic and ethnographic fieldwork have revealed the existence of several ethnic or historical subregions, and Mindeleff himself was aware of several distinct kinds of settlement patterns. Yet the overall uniformity of prehistoric house-types is still recognized as a significant characteristic of the whole landscape.

How is such uniformity to be explained? In the nineteenth century scholars assumed that it was the result of environmental factors: that identical or similar climates, topography, and natural resources produced identical or similar kinds of dwellings. But there are also social factors to take into account. When we conform to local ideas as to the role of the dwelling and its relationship with neighbors and with the place of work, we are in fact conforming to the local house-type. The Pueblo communities of the prehistoric Southwest, like most isolated and autonomous farm villages, attached great value to homogeneity. All observed much the same public rites and ceremonies, lived by the same fixed agricultural and religious calendar, and all were dedicated to preserving the same intricate social order. In consequence, the dwellings of the prehistoric Pueblo Indians were similar to one another, not only in each individual village but throughout the region.

A standard type of vernacular dwelling, primarily for working people, is a fairly common phenomenon, and indeed becoming even more common as public authorities undertake to provide dwellings on a wholesale scale. But what we are likely to find impressive about the southwestern version is that its uniformity rarely suggests enforced compliance, or even a conscious desire to conform. No matter how similar they may all be as to shape and size and construction, the prehistoric Pueblo dwellings somehow manage to tell us that each is an individual achievement—the private, almost instinctive expression of what must be an ancient prototype. To the European-American student, Pueblo architecture is extremely hard to understand. The prehistoric dwelling is *not* architecture, and it is modest enough in scope to invite speculation. Unlike the larger and more imposing prehistoric collective structures, it allows us to compare our own ways of building with those of a vanished culture. One mystery confronted can lead to the recognition of others.

It is generally agreed that the basic unit in Pueblo architecture is the room. Since the width of a room is determined by the length of the roof beams, we rarely find a room much wider than fourteen feet, and a room of twelve by twenty feet would be accounted large. Many rooms, possibly used for storage, are no more than five feet square. When it is

Taos Pueblo, New Mexico. (Author's photo)

said that this boxlike room or cell is the basic structural unit, we mean that no larger or more complex unit of construction was ever devised, and indeed the room is the equivalent of the dwelling in its simplest form. But paradoxically enough, this basic unit is never self-sufficient. It must always be related to a row or collection of similar units, which is perhaps the equivalent of saying that the *dwelling* as the domicile of the nuclear family is never self-sufficient but viable only when it is used in conjunction with other domiciles, other rooms occupied by members of the extended family or clan, all centered on a family shrine.

The second implication of this definition of the room as basic unit is this: Every Pueblo building, no matter what its size, is actually a *cluster* of such cells. Even the largest community house, like the one at Taos, six stories high and a quarter of a mile long, is a honeycomb of innumerable small rectangular rooms. It may be objected that this is also true of every modern European-American building: are they not all essentially clusters of cells of varying dimensions? Yes, but they are designed and constructed as a unit subsequently divided. With us the frame, the armature comes first; with the prehistoric Pueblo builder the cell or room came first and was then duplicated.

One of the most valuable insights into the nature of Pueblo architecture is Benjamin Lee Whorf's comment that the Hopi language contains no word for *room*. "We are struck by the absence of terms for interior three-dimensional spaces, such as our words 'room, chamber, hall, passage' . . . in spite of the fact that Hopi buildings are frequently divided into several rooms, sometimes specialized for different occupancies." Is it possible that we, no less than the Hopis, are unconscious victims of our language when we attempt to interpret architecture? We are almost incapable of perceiving the dwelling as other than a collection of rooms. We even speak of a one-room house, as if the room were somehow distinct from the interior of the house. The Hopi see the house as a receptacle; its interior is temporarily defined in terms of its temporary content. "The occupancy of a piki-house is called by a term which means 'place where the griddle is set up' but it would be called that if set up outdoors, and there is no term for the piki-house itself."[2]

A Pueblo room (or basic dwelling) is thus nothing more than a three-dimensional interior space in which objects are contained or events occur. The room itself imposes no identity on its temporary content, and in turn the contents do not permanently characterize the room. Only a collection, a sequence, of rooms can have that effect. It is as if the occupants were saying that the single space, the single event is of no consequence: it is *repetition* which creates the periodic or rhythmic recurrence of spaces and events, the cosmic order.

The rectangular, flat-roofed building, call it dwelling or room or receptacle, is made of stone or adobe or a combination of the two. We tend to associate it with adobe, but there is good evidence that adobe was preferred by Indians in the Rio Grande Valley. Some authorities believe that several prehistoric North American communities knew how to form adobe into something like bricks. But in the great majority of structures the adobe was simply compacted into balls or wads and put in place, course by course. (The so-called puddling technique is described in more detail in the following chapter.) Walls thus constructed were likely to be fragile and easily eroded by rain. The stone walls were frequently laid up without mortar, smaller stones being wedged into the cracks and spaces. Adobe was also used to close the chinks, but this could occur only after a rain had provided water to make mud.

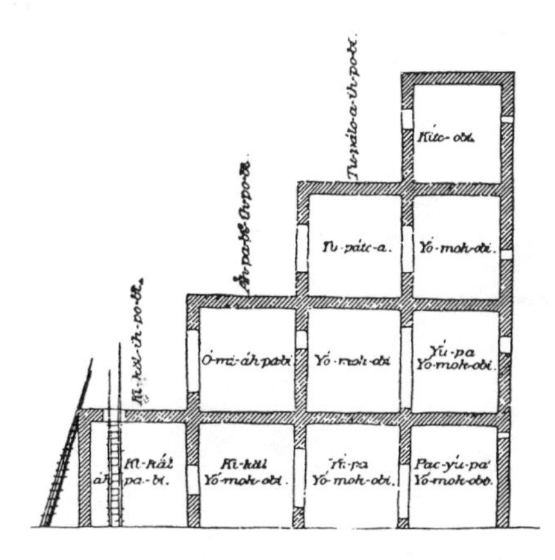

Cross-section of multistoried Pueblo dwelling unit. (From Victor Mindeleff, *A Study of Pueblo Architecture in Tusayan and Cibola,* 1891, reissued Washington, D.C., 1989)

The typical foundation is rarely excavated and usually consists of a row of rocks and stones posed on the surface of the site. The roof beams—logs of aspen, spruce, or pine brought with some labor from the nearest forest—are peeled, but never dressed or squared. A peculiarity of the dwelling is that it has no door leading to the outside on the ground floor, but is entered by means of a trapdoor in the roof, and a ladder. The effect must be to transform the ceiling into a surface connected with the world; not a distinct element in the room but a continuation of the walls on another plane: the sense of containment, or stereometric space, is reinforced.

All of these materials are the product of the local environment—a spatially very restricted environment in that the Pueblo Indians had no beasts of burden. To quote again from Cosmos Mindeleff: "One of the peculiarities of Pueblo architecture is that its results were obtained always by the employment of materials immediately at hand. In the whole Pueblo region no instance is known where the material (other than timber) was transported to any distance; on the contrary, it was usually obtained within a few feet of the site where it was used. Hence it comes about that difference in character of masonry is often only a difference of material."[3] In this connection, therefore, "available" means not only those materials that are easy to carry but those that need little or no processing—which is to say that they are not necessarily the strongest or most adaptable materials nor those that last the longest. Whoever has examined Pueblo craftsmanship must be aware that the builders were

Houses built over irregular sites, Walpi. (From Mindeleff, *A Study of Pueblo Architecture*)

well acquainted with the properties of the various materials and were capable of extremely precise and beautiful results, particularly in their use of stone. Many of the walls they build show considerable preparation: chcosing uniform size and texture, laying bands of distinct color, and chinking with small stones often produce what must be a deliberate artistic effect. Indeed, it is a concern for smooth and uniform textures that leads the Pueblo builder often to cover a rough stone wall with a coat of adobe.

What one eventually notices, however, is the lack of concern for establishing any bond with the immediate environment—more specifically, the lack of awareness of the natural forces or processes that will in time threaten the security of the building. It is well known that the Pueblos builders did not use the keystone, arch, or column, and that they used the buttress with little skill. They often did not build even the slightest foundation. The corners in their masonry houses were never bonded, and a wall supporting three or four stories is often no sturdier than one which supports the roof and roof beams only. Experience must have taught them that thrust and weight, settling and expansion ought to be taken into account, but it was a lesson which they chose to ignore. Why such skillful and ingenious builders should so choose is one of the mysteries of Pueblo culture; one can only surmise that there were some aspects of the natural environment, some forces within it, that they could not or would not recognize. "The absence of any attempt to improve the natural advantages of the site is remarkable," Mindeleff remarks. "No expedients were employed to make access either easier or more difficult, except that here and there series of hand and footholes have been pecked in the rock. . . . The cavities in which the [Canyon de Chelly] ruins occur are always natural; they are never enlarged or curtailed or altered in the slightest degree, and very rarely is the cavity itself treated as a room, although there are some excellent sites for such treatment." Elsewhere he describes the futile attempts of the builders to use timbers for the foundations of walls.[4]

The world offers us two different ways of interpreting the passage of time and of ordering the time we live by, and here in the Southwest both of them derive from the beauty and immensity of the region. The first way is by observing the annual movements of the heavenly bodies and how those movements create the sequence of days and seasons and the four directions. From the predictable recurrence of summer and winter, night and day, life and death, from the cyclical nature of celestial time we learn to devise a formal social order, an elaborate calendar, and a harmonious way of life that promises to endure forever.

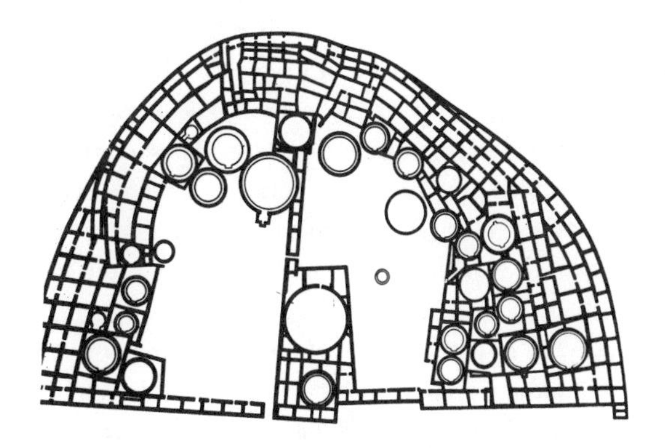

Plan of Bonito Pueblo, Chaco Canyon.

The other way of interpreting time comes from our everyday experience of the world around us, from contact with the earth itself. Here is where we are confronted with uncertainty and constant, unpredictable, irreversible change, which we are powerless to oppose. Whether the events that form and transform our environment are sudden and brief like a flood or the crashing down of a canyon wall, or whether they last for decades like a drought, their time scale teaches us nothing. Nevertheless, it is possible to remember them; in the course of generations we can understand this earthly system of time and eventually eliminate some of its terrifying and destructive unpredictability.

The prehistoric Pueblo people seem to have chosen that celestial, antihistorical concept of time, and to have given little thought to that other time which ticks off its passage by slow erosion or the widening of a crack in a wall or the crumbling of a foundation. They built as if the present order were going to last, untroubled by age and neglect and decay. Benjamin Whorf's exploration of the Hopi concept of time and his essay on Hopi architectural terms are not only a brilliant revelation of the connection between language and notions of time, but a revelation of how our architecture differs from that of the Pueblo Indians.

One radical difference is our European-American awareness of history. Any architectural student is likely to conclude that the vernacular dwelling of prehistoric and Dark Age Europe had certain similarities with that of the prehistoric pueblos: both types were basic all-purpose rooms, both were receptacles, both were built out of locally available materials, and both were examples of tectonic structures; last, according to archaeologists on both sides of the Atlantic, neither house-type evolved in any significant manner during a period of more than five hundred years.

But (again according to European archaeologists) the vernacular or peasant dwelling in northwestern Europe began, sometime in the tenth and eleventh centuries, to undergo a series of changes, largely in construction. What these changes seem to have signified is that the European dwelling ceased to be a receptacle, a container, and evolved as a distinct organism, with interior features having no direct connection with the exterior: new types of partitions, new forms of framing and joining, the procedure of building in terms of bays, roofs, and walls that resisted the weather more effectively, better foundations that resisted the action of frost, better lighting, better and easier access—all became features of

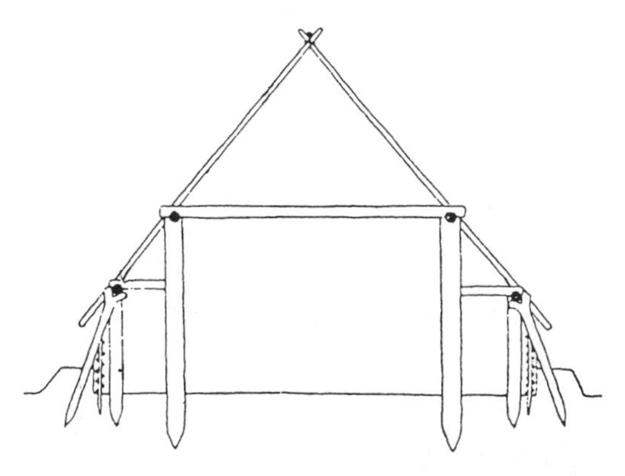

The dwelling as counterenvironment. Cross-section of early medieval northern European dwelling. (From Jean Chapelet and Robert Fossier, *Le Village et la maison au Moyen Age,* Paris, 1980)

even the simplest dwelling. Some of these improvements were imported from elsewhere; others resulted from the use of better tools or better building materials. The upshot was that the vernacular house in medieval Europe began to resemble what might be called a *counterenvironment:* an environment (or microenvironment) designed to resist the assaults of the world outside and above and beneath. The dwelling became an instrument for production and for the preservation of the family: the interior was seen as distinct from the exterior. The house became a complex and flexible artifact, largely because it established a new and effective identity as a counterenvironment.

The Pueblo dwelling never did this; perhaps because of a more equitable climate, but perhaps also because of a deep-rooted reliance on the house or room as the minimal structure, its use and form determined more by social than by environmental factors. If it was a question of giving a name to this type of dwelling, it could be termed *proto-vernacular*—not yet completely vernacular in its dependence on a racial archetype and its lack of a clear-cut distinction, structurally speaking, between exterior form and interior space. It may well be that as we study the prehistoric Pueblo dwelling in more detail we will discover the beginnings of the full-fledged interior, the counterenvironment. Under the influence of European-American examples the contemporary Pueblo dwelling has already become vernacular.

4 Church or Plaza?

Church interior, Trampas, New Mexico, c. 1935. (Courtesy Museum of New Mexico, Santa Fe. Photo: T. Harmon Parkhurst)

In a few years' time we will be marking the four-hundredth anniversary of the first Spanish settlement in New Mexico. It took place in July 1598, and the founder of the settlement was Don Juan de Oñate. A year earlier he had set out from Mexico as leader of an expedition to explore and take possession of the largely unknown northern frontier region, and to convert its Indian inhabitants to Christianity. Together with a small party of officers he went on ahead of the main party and traveled north up the valley of the Rio Grande until he reached a landscape suited to colonization. Here the group discovered an abandoned Indian village. Oñate chose this as the place to halt; in a sense this was the goal of the expedition.

A brief ceremony ensued. Mass was said and Oñate solemnly declared the village to be the future capital of the province of New Mexico, of which he was to be governor. Giving the village the name of San Juan de los Caballeros, he brought the ceremony to a close by planting a standard specially designed in Mexico for the occasion: a white silk banner bearing the images of the Virgin and Saint John the Baptist. It was adorned with tassels of gold and crimson.

Ironically enough, the capital of the province was soon moved to another place, and Oñate, captain general and governor for life, fell into disgrace and was removed from office. Yet the short, largely improvised ritual in the wilderness, witnessed by no more than a scattering of silent Indians, survived (and still survives) as a Spanish New Mexico tradition.

The custom of celebrating as formally and splendidly and as publicly as possible a significant historical event was popular in sixteenth- and seventeenth-century Europe, as a visible expression of a widely held belief that history was a preordained sequence of trials and triumphs, a stately procession of events leading to a transcendent climax. Entirely alien to the thinking of the Native American population, it took root among the colonists as a symbol of their Old World heritage, civic as well as religious, and as an effective way of celebrating their accomplishments.

Other, more elaborate celebrations of the Conquest soon followed, each a well-staged display of the hierarchical social order, each marking, so it was hoped, a further advance toward the creation of a Baroque commonwealth in New Mexico. There was a celebration—procession, mass, sermon—when the main body of the expedition finally arrived; a celebration on the completion of the first irrigation ditch, a celebration of the building of the first church, and a three-day celebration on the occasion of the public submission of local Pueblo chieftains to the authority of the king of Spain. It included a mock battle between the Moors (on horseback) and the Christians (on foot, though equipped with firearms). A sermon terminated the event.

The euphoria of conquest soon evaporated, for the province of New Mexico fell on evil days. There was rebellion in some pueblos, mutiny among the soldiers, and many settlers fled to the safety of Mexico. Disorder and discouragement were the result, and the modern landscape of New Mexico contains few visible reminders of the first two centuries of colonizing.

Those few are likely to be the remote rural churches built by the pioneer Franciscan missionaries. Church architecture thus provides us with the most reliable evidence of continuity in much of New Mexico's history. In his book *Sanctuaries of Spanish New Mexico* (Berkeley: University of California Press, 1993), Marc Treib analyzed the origin, construction, and subsequent modification of a number of those early colonial churches. Those which he described in detail are presented both as religious symbols and as the products of a frontier economy and of Indian labor; interpreted as it were as provincial variations on Mexican Baroque and as intrusions on the ancient Indian way of life. Treib's breadth of historical and architectural learning is paradoxically best revealed in his discussion of prosaic details: the limited choice of materials, the limited choice of tools, the restraints imposed by soil and climate. The merging of two very different perspectives gives the reader a stereoscopic view: the church as artifact, separated from the Pueblo villages by its European complexity; but also the church as an integral part of the economy and society of a workaday Spanish colonial landscape.

Most of us are aware that the prehistoric pueblos—at least those in the basin of the Rio Grande—were built out of adobe; and anyone acquainted with the Southwest knows that the Spanish conquerors introduced the use of the adobe brick. Few writers on southwestern architecture have said much about the contrast between Pueblo construction techniques and those of the Spaniards. It is generally assumed that the transition from one to the other was smooth, when in fact it took several generations. Treib provides us with a concise description of the two ways of using adobe, and by simply enumerating the characteristics of each he alerts the reader to an important fact: building with standardized adobe bricks—which is what the missionary church builders did—represented a totally new approach to architecture in the eyes of the Pueblo Indians.

Although pre-Columbian cultures of South America knew how to make and use adobe brick of standard dimensions, the prehistoric Pueblo adobe building technique was an example of what is known as puddling. It involves taking handfuls of mud (without any binding ingredient like grass or straw) and squeezing them into lumps or balls, then placing these lumps in rows or courses, one after another, up to a convenient height. This produces a very thin wall without any reinforcing frame; moreover, there seems to have been little or no preparation of the site—to say nothing of building a foundation for the house. As Treib notes, the puddling technique closely resembled the way the Pueblo Indian women made pots: by using rolled coils of clay. Both activities were, in fact, considered the occupation exclusively of women.

A puddled wall was rarely sturdy enough to support a second story, and yet second stories on top of puddled houses seem to have been common. How many structures collapsed is anyone's guess, but there is ample evidence that they often *did* collapse. It is surprising that no method of reinforcing puddled walls was ever devised, for Pueblo builders frequently showed great skill and ingenuity in masonry. All that we can assume is that two-story puddled houses were not thought to deserve much investment in labor, and were not considered permanent. They were easy to build and easy to abandon.

The Spanish invaders brought with them from Mexico (and ultimately from Spain) the knowledge of how to make strong, uniform adobe bricks, and of how to use them in construction. This was much more than a primitive or peasant building material. Ordinances drawn up in Mexico City in 1599 called for the licensing of adobe masons, who were required to know how to build foundations, how to calculate the weight

Prehistoric pueblo masonry, emphasizing smooth surface rather than strength. (From David Muench, *Anasazi,* New York, 1974)

of the roof in relation to the thickness of the walls, how to erect scaffolding, and even how to read plans. Whether these regulations were ever enforced in New Mexico is doubtful, but their memory persisted among the priests and soldiers and settlers, and the prototypal building—church or dwelling—which they sought to reproduce on the frontier was one with strong foundations, sturdy walls supporting heavy roof beams, and a number of precautions against settling and collapsing and eroding.

To the Indians who did the heavy work, much of the building activity must have been bewildering: the use of plumb lines and measuring rods, and the frequent consulting of a plan or drawing. But one thing must have been obvious to them all: that the design of the church and all the work done in it were intended to achieve a single purpose—the creation and protection of a number of interior spaces. This meant building a structure so strong, so massive, so firmly attached to its site that it could resist the ravages of time and weather; a building which could last for several generations.

The use of adobe bricks and the use of puddling are so totally unlike in every respect that there seems no need to compare the two. Yet the average Pueblo Indian of the colonial period probably had no trouble in seeing the essential difference: the house of adobe bricks was meant to be the most important space—or collection of spaces—in the daily existence of its occupants, a substitute for life outdoors; whereas the house of puddled mud was simply a useful adjunct, designed to hold things or to serve as part-time shelter. The church was a particularly fearsome example of adobe brick architecture, not only because of its monstrous size and permanence, but also because people were compelled to enter its labyrinthine rooms at regular intervals and for a prescribed length of time. Treib quotes from priestly regulations calling for guards inside the church door to prevent Indians from leaving before the end of the service.

Once entrapped in the church, the outsider became a helpless participant in the elaborate, highly organized pageant progressing down the center of the nave to the altar. The other-worldly atmosphere of the interior was enriched by darkness, music, incense, and the costumes of the officiating priest. To the European members of the congregation—priests and soldiers and settlers—this was not only very familiar, it was also a reminder of another well-established public event, the Baroque celebration of earlier and happier times: far smaller, far less splendid, far less varied in composition, but still retaining the basic elements: symbols and hierarchy and of course the slow procession, building up to a dramatic climax at the altar. Both kinds of ceremony illustrated the Western concept of the interaction of time and space; progress, movement in either dimension inevitably led toward a wished-for goal.

How did the Indian participant respond? The bells, the measured tread of the processioners, the rhythmic order of the service itself, as well as the reminders of the church calendar culminating in the observance of Easter, all produced an ever-increasing expectation of the final moment. But it was not the time the Indians lived by: that was a cyclical, never-ending recurrence of cosmic events, the movement of heavenly bodies, the sequence of seasons and of night and day; ordering of time of the outdoor world, not of the world of dark interiors. To escape from the enclosed spaces of the church was to escape from the tyranny of an incomprehensible architecture and to return to traditional ways of thinking.

As I mentioned in the preceding chapter, Benjamin Lee Whorf suggests that the Hopi (and presumably other Pueblo Indian societies) had an ambivalent attitude toward all interior three-dimensional spaces. He writes that though the Hopi had terms for many architectural structures defined by their outward appearance, they had no term for the room.[1] This tendency to perceive the room or any other interior space (such as the nave of the church) simply as a container with no inherent quality or function of its own is consistent with the use of the flimsy, essentially short-lived puddled dwelling. For the Pueblo Indian, the really essential lived-in spaces are those found in the village: the plazas and alleys and gardens.

An unusual element in Treib's study of the missionary churches is his account of the decay among many when their missionary role was reduced or eliminated. It was then that they acquired an archaeological appeal, and Treib's discussion of preservation and restoration policies emphasizes the dangers residing in a too-strict attempt to restore the churches to something like their original condition. Much of the "incorrect" or inaccurate restoration in the nineteenth century was the work of the Pueblo Indians themselves, and it indicated how their attitudes toward architectural spaces had changed. More than a century ago Victor Mindeleff, in his study of Pueblo architecture, noted that the Zuni Indians had attempted to patch up their decaying churches by using adobe, but

Zuni dance in the pueblo plaza. (Courtesy Museum of New Mexico, Santa Fe)

with little understanding of the nature of adobe bricks. "When molded adobe bricks have been used by the Zuni, in house dwellings, they have been made from the raw material just as it was taken from the fields. As a result these bricks have none of the durability of the Spanish work." Mindeleff added that "walls in Zuni were only as thick as necessary . . . evidently modelled directly after the walls of stone masonry which had already been pushed to the limit of thinness."[2] It was as if the nineteenth-century Pueblo builders were reverting to the short-lived construction of pre-Conquest times.

If their neglect of the church buildings had any deeper significance, it was that the Pueblo people were turning away from the built environment in which they had never felt at home. It was then that the ceremonies they most cherished—the traditional dances, endlessly repetitive and without climax, whether in time or in the space of the plaza—were once again freely honored, and performed in the open.

A traditional church, locally designed and built. San Isidro del Sur, New Mexico. (Author's photo)

Those who believe in the persistence of a Baroque heritage among the Spanish-American population of New Mexico can take heart in the survival of many church traditions. The cultural, as distinguished from the doctrinal, influence of the Catholic church is particularly strong in northern, predominantly rural counties. Despite a dwindling population, increasing poverty, and an omnipresent Anglo culture, there are still villages which look upon the church and its priest as defenders of a formal Spanish way of life. It is in the church that they expect to hear correct Spanish and to observe correct behavior and dress. It is in the church that they celebrate marriages and baptisms, and where they gather to mourn a death. There are few other occasions for honoring ancestral traditions and family ties. But on the day (or the eve) of the local patron saint, something like the ceremonious procession reappears. The church bell rings, there are bonfires at intervals around the outside of the church; the image of the saint heads the procession as the congregation walks slowly three times around the church, singing as it goes. The flames light up the rough adobe walls and the earnest faces. All the elements of celebration are present: all the symbols of order and reverence and undying love of this particular time, this particular place.

5 The Mobile Home on the Range

Trailers in the New Mexico landscape. (Author's photo)

New Mexico contains an extraordinary variety of dwellings; I doubt if any other state has as many. We have Pueblo Indian, Spanish-American, Anglo-American, and Navajo houses. Some ancient house-types are eight or nine hundred years old, and how they were used is still not entirely understood. Chaco Canyon, the largest archaeological site in the United States, is from that point of view the greatest of mysteries. But we have house-types like the trailer (or mobile home) that are new and evolving, and the brief counterculture of the 1960s also made its contribution.

What is unusual about many of these house-types is that they can be found side by side in the same small community; and that in many cases they are lived in by the families who originally built them. This means that it is easy to find out how they were built and how they are used. To be sure, there are many traditional house-types in the East and in the South which have been studied and preserved. But when they are old they have often had so many owners and have gone through so many remodelings that the link between the original builders and the houses as they now stand is hard to fathom. As a result, the architectural historian or preservationist has to confine his or her work to investigating the house as artifact—materials, tools, and methods of building. Thanks to sophisticated archaeological techniques, we have learned much about old construction processes, and about the ethnic or geographical origins of the builders, and we can, in fact, restore the whole structure to something like its original condition. Yet we can only speculate about how the house was used in daily life, how it was related to its working landscape. New Mexico thus has something of value to offer in the way of insight.

Most dwellings in New Mexico are small and as works of art or architecture hardly worth close examination. Nevertheless, they give our landscape its character. Usually they are of simple construction, not meant to last for long, for they are likely to be homes of families with little money—small-farmers, wage earners, day laborers, and men and women working in service jobs—people who often think of moving in hard times to where there are better opportunities. If they can rent or sell the house they are lucky; if not, they usually abandon it. It is true that New Mexico has a number of handsome architect-designed houses in Albuquerque and Santa Fe, but in my travels through the less prosperous parts of the state I have seen village after village, and even small towns, where there is no such thing as a mansion, a Great House—one costing money and clearly adapted to a different lifestyle.

What I have seen almost everywhere are small, unpretentious one-story houses, sometimes clustered together, sometimes strung out along both sides of the road, sometimes sitting alone in the open country: in most cases a sort of basic, all-purpose house-type for a family with an income below the national average and with an above-the-national-average number of children and used cars.

When I first came to New Mexico in the 1920s, I was attracted by the Spanish-American villages scattered throughout the ranch country where I was staying. They seemed very foreign, very un-American. In those times most families supported themselves by farming and raising cattle or sheep. It was a hard life; many men worked as sheepherders or cowhands on ranches in Wyoming or Colorado and were away from home for months at a time.

The villages were half-hidden in the immense open rangeland, near a stream that watered the small fields of corn and chili and beans. The surrounding landscape was organized in an almost medieval manner. Easterners are not always aware that communal control of the land and its use, with a large common for the livestock, existed in the Spanish Southwest before New England had been heard of. In some places the system still survives. Each household had the right to graze its cows or sheep on the very extensive community range, to use water from the community irrigation ditch, and to take wood from the community forest. Furthermore, the church belonged to the villagers, and so did the few roads and trails. Whoever used those facilities had to help keep them up, and this sharing of work and land and responsibility helps explain the strong sense of local solidarity among the villagers. This kind of community, harking back to the late sixteenth century and brought from Spain by way of Mexico, had already begun to fall apart by the time I arrived in New Mexico. Sizeable private holdings had come into existence, and community obligations were more and more neglected. Younger men and women left to find work in town. But certain customs persisted:

Traditional Spanish-American village architecture. (Author's drawings)

The community cattle roundup and rodeo were popular events, and so was the old-fashioned public church celebration of Christmas and Easter; everyone still went to the village church.

Architecturally speaking, the houses were far from remarkable: one-story structures with two or three rooms, usually of adobe. Often they had a pitched roof of corrugated tin that shone in the sun, and a long front porch. They were all very much alike, for there is a limit to the variety that can be introduced into the plan of a house with two or three rooms. None had running water or electricity, and almost all had dirt floors. They were painted different colors, however; bright green or pink or brown, with white window and door trim, and entirely without ornamentation. The manner in which they were sometimes connected in rows to form three sides of a common courtyard or plaza gave the village an almost urban aspect. Throughout the day the houses were quiet. They were scantily furnished, yet their interiors gave the impression less of poverty than of an austere formality. When I visited such a house I went not to the front door (painted white and locked) but to the kitchen door, and (as was the custom among neighbors) I entered without knocking. The mother-in-law, babysitting, said nothing by way of greeting. I asked where Manuel was. She answered that he was out, getting a load of firewood. And Joe? "Joe is out seeing about a job," she said, adding that Linda was also out, having gone to the store.

This was almost always the case: at every house everybody was out; being "out" meant taking some part in the life of the village. Early each morning the man of the house rode out to work in one of his fields or to check on his livestock grazing in the common. He kept an eye on what there was that he could use: a dead cedar tree for a fence post, a stray hen, a rare medicinal herb. A passerby told him of a rancher who needed extra hands for putting up hay or building a fence. He stopped at the store to get a sense of how his credit stood. Every item, every contact was of possible worth. It seemed to me that the men in the village were always aware that they were involved in sharing the land: the water, the grass, the sand and gravel, the game in the forest. They were always on the lookout for some abuse, some reason for holding a noisy public meeting and clamoring for justice. And this incessant clamor suggested not that justice was hard to get but that it could be had on demand. We who live in town think of the countryside as where people farm or enjoy the beauties of nature. Actually, it can be a stimulating place, and politically speaking even the most somnolent village has much to offer, for that is where we see custom in action, regulating movement and ways of work and relationships between neighbors. It is where we eventually recognize that an established order is not easily changed. Remaining at home would not only be lonely, it would mean that you were deprived of the excitement of community existence and its opportunities.

In the pastoral New Mexico of more than half a century ago, I believe the distinction between the quiet domestic realm and the community was very clear-cut, and that the community mattered more than the house. Certainly the house was important; it stood for shelter and privacy, it was where family ritual was enacted, year after year. It was the place of origin and the place where you died. But it was remote. Its economic role was limited. It was not the place of work in the sense of being the place of gainful employment. It was not the place for conviviality, and it was not a personal work of art to be proud of. Indeed, the reason all houses resembled one another was that every man was capable of building one. Finally, the house was rarely associated with memorable family events: aside from religious images, it contained neither mementos of the past nor provisions for the family future. Marriages and funerals and anniversaries and reunions were held in the church or the school, or sometimes in the local dance hall.

I realize that even the simplest dwelling demands respect for its rich symbolism and the memories it holds. I can only say that it has another, more prosaic aspect: the house as a space or a composition of spaces and walks and doors that makes certain relationships possible and impedes others. I had the sense that the village houses served as supplements to the larger, shared spaces of the community, incomplete in themselves, fragments of a much more complex unit. The typical house did not pretend to be anything more than simply one house among several. Its role was to make visible how the inside world related to the outside, how the individual related to the village, and how the hours of working with others were distinct from the private routine of the home. If I had been asked to define a vernacular dwelling, I would have thought of those Spanish-American village houses and said that it was a house which depended on its immediate environment for the satisfaction of daily needs.

Sixty years, however, is a respectable length of time, and when I set out in 1990 to refresh my memory of the past, I found many changes in the New Mexico landscape, the most striking being what had happened to some of the villages I had once known. They had degenerated into rural slums of a very abject kind. Fields had reverted to second growth, houses were in decay, roads and irrigation ditches choked with rubbish and abandoned cars. I cannot account for this tragic decline, except to say that New Mexico is always subject to entropy. Farming no longer provided a living, most villages were too isolated, too lacking in resources, to try another way of life. So now, twenty miles from Santa

Fe, with its opera, its dozens of art galleries, its polo fields, you can find a squalor as hopeless as anywhere else in the United States.

Yet, some of the villages *have* survived. The men now work for wages in the service sector or do odd jobs in town. Thanks to the automobile, the environment they depend on has expanded well beyond the village and ranch, and they think nothing of commuting thirty or forty miles to work. The roads are paved and buses take the children to a consolidated school. When they come home in the afternoon and run down the street, their bright clothes and loud voices bring life to the village and mark the time of day. Every household seems to have at least three cars, one of them a pickup. Cars in varying stages of mobility are parked outside the bar, the convenience store, in front yards, in deserted corrals, and in vacant lots. With hoods raised, they seem about to devour the young men adjusting the carburetor, and Spanish music comes from their radios. Though the villages are probably just as poor, comparatively speaking, as they were in the past, some now have more movement and more vitality.

I was struck by the number of new dwellings and the decay and abandonment of the old. Adobe is no longer the most popular building material; cement block and frame houses are common, as well as houses trucked in from elsewhere. The new ones are more comfortable than the old ones ever were, with electricity and gas and running water. They indicate that a number of the men in the village have been "in construction" (as the phrase goes) and have learned about new materials, new ways of building, and how to use power tools.

A great deal of this new housing throughout New Mexico—and for that matter throughout the whole country—consists of trailers. They are everywhere: tucked in between houses, attached to houses, even on top of houses; in alleys and gardens and out in the fields. In fact, there are New Mexico villages where trailers outnumber conventional dwellings and where the newly arrived tourist cries out in delight at the glimpse of an adobe house. I was sorry to see that the old close relationship between houses, suggesting as it did a relationship between members of the same extended family, had been replaced by a more scattered arrangement, like that among friendly but self-sufficient neighbors. The newer, freestanding houses seem to prefer the margins of the road leading out of the village to the traditional compact pattern of plazas.

It was the sight of those trailers that made me question the accepted definition of the vernacular. Like most Americans of my genera-

House combined with trailer. (Author's photo)

tion, I first became aware of trailers in the years during and immediately after World War II, when they were widely used as a form of emergency housing. They clustered by the hundreds around Army posts and construction sites, and after the war they invaded college campuses to accommodate married students. They were unsightly, but at the time no one cared. They were temporary, and in fact most of the wartime trailer communities have long since vanished.

Now, two generations later, America has more trailers than ever before. Called mobile homes in the trade, they are larger, more comfortable, and more expensive. More than thirteen million Americans, most of them from young blue-collar families, live in trailers and (for the time being, at least) call them home. There is, however, strong public prejudice against them. Architecture historians have learned to accept the bungalow, the split-level ranch house, and the A-frame, but still they cannot bring themselves to recognize the trailer as a dwelling. Few property owners want trailers in their neighborhood, and style-conscious communities do what they can to relegate them to the less visible parts of town. Trailer parks are often researched by sociologists, who report on the deadly conformity among the tenants or on the tyrannical behavior of the trailer park managers. Yet it is one thing to glimpse in passing the regimentation of a trailer park, and quite another to see the solitary trailer permanently part of the fabric of a neighborhood. Far from being hidden or disguised, it is often conspicuous.

Over the years the educated public, led by architects and urban planners, has drawn up the indictment of the trailer. It is part aesthetic judgment, part structural critique, with a touch of compassion for those who are unfortunate enough to have to live in one. To begin with, the trailer is an industrial product, mass-produced, low-cost, and disposable. It comes out of a midwestern factory and is shipped by truck, quickly unloaded, and soon ready for occupancy. It has bypassed the craftsman and the architect and the landscape architect, and the owner (or consumer) has no opportunity for self-expression, or even a say in the ordering of the interior or in the outside decorations. Some trailers come completely furnished—the ultimate in standardization. And then, coming as it does off an assembly line, the trailer ignores local architectural traditions and local environmental constraints. Its uncompromising shape and boxlike appearance make any real composition of a group of trailers impossible. No matter how we site them in relation to one another, we never achieve anything like a traditional village.

For all its conspicuous bulk, the trailer is quite small and cramped, and, long and narrow as it is, fails to provide a half-way satisfactory arrangement of rooms for a family. Though it is often efficient as a store or an office or a schoolroom, its inflexibility means that it can never be a self-sufficient, autonomous dwelling. On the contrary, almost from its first day of occupancy it spills its contents—and its occupants—into its surroundings: parked cars, refrigerators, packing cases, children and dogs and laundry invade the landscape. As time goes on the trailer becomes more and more dependent on the spaces provided by village taxpayers: cars take up room, children need spaces for play. Yet, ironically enough, the trailer rejects assimilation: its potential mobility, its frequent changes in occupancy and ownership, its ambiguous legal status all work against its acceptance. Finally, it quickly becomes shabby in appearance. It is of light construction, easily destroyed by fire or toppled by a high wind. Literally as well as figuratively, the trailer has no real attachment to place.

Most would agree that these are valid criticisms; we could probably add to them. But from the point of view of those who live in trailers I think they miss their mark. From what I have learned, the villagers who have moved into trailers are in general satisfied. They wish their trailer were larger and had better insulation. They object to the floor plan. Nevertheless, to them the advantages of the trailer far outweigh

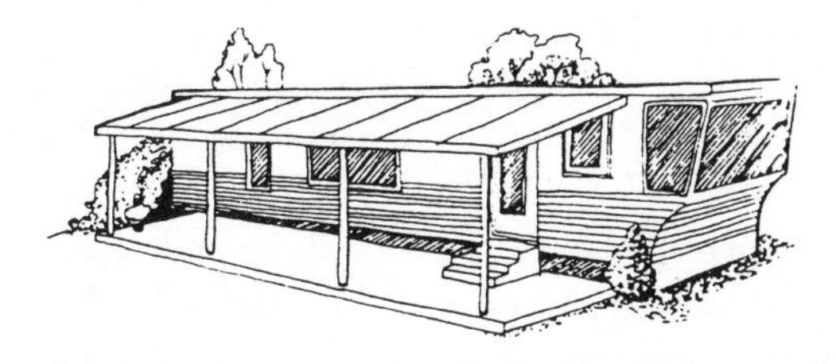

Mobile home designs, 1970s. (From A. D. Wallis, *Wheel Estate,* New York, 1991)

its faults. What they especially appreciate is how little the trailer costs, compared to even the smallest house, and how easy it is to finance. They regretted leaving the old adobe house with its associations, but it was a joy to move into a brand-new home, clean and never used.

Newness is something we do not always appreciate, but I am convinced that a large minority of Americans have never owned a new car, though they would like very much to. That is why there are spray cans to produce the smell—whatever it may be—of a new car interior. A new trailer has the same exciting appeal: stickers on the windows, books of instructions, and that indefinable smell of newness. It takes only a few days to realize how convenient and comfortable the trailer is, and how easy to maintain. The fact that it resembles all the other trailers in the vicinity is if anything a source of reassurance, for it means that the choice was a popular one, endorsed by other families. The most welcome feature of trailer living for the villager is that it brings with it no new responsibilities, no change or expansion in the traditional domestic routine. Nor does it alter the old relationship with the outside world: the man or woman of the family can as usual leave home in the morning, only with the trailer there is no chopping of wood, no feeding of livestock. Life is simplified and begins, as it always has, when we join others in work and conversation. Moreover, with fewer domestic chores the wife is at last free to move into the community. Trailers, as we all know, are rarely mobile in the literal sense of the word, but because of their impersonality, their fungibility, they are like automobiles: easy to trade and sell. When a better job becomes available somewhere else, the family can at least consider the wisdom of selling, and of finding similar accommodations wherever they go.

I need not belabor the point: for a great many families the trailer is a sensible way of living. Indeed, it almost seems as if those shortcomings which critics never tire of mentioning—the lack of individuality, the functional incompleteness, the dependence on outside services and amenities, and even the lack of such traditional architectural qualities as firmness, commodity, and delight—all are what make the trailer useful and attractive to many of its occupants.

I am no blind admirer of the trailer or mobile home. I have seen at first hand what is wrong about its plan and construction. But I still think it is the most practical low-cost dwelling we have, and that it is well adapted to a way of life that is becoming increasingly common in both urban and rural America. That way of life is identified with the blue-collar worker: the man or woman without capital, without any marketable skill, and with only a limited formal education. The man or woman of the family (in many cases both) has to work by the hour or the day at an unskilled or semiskilled job away from home, with little assurance that it will last. These factors obviously have their effect on the kind of house they can afford, and how they use it.

There was a time when we would have used the term *proletarian* to describe this class of citizen. As long ago as sixteenth-century England, the word signified a worker who owned no land and who could produce

The trailer can never be a self-sufficient, autonomous dwelling. (Author's photo)

nothing for the market. But Marxist theory has redefined "proletarian" to exclude the farmworker and craftsman, and it now means the industrial worker alone; and of course the industrial worker is often highly skilled and well paid. So if we want a convenient term which would include all those who work in the less prestigious service jobs, we should probably say "unskilled wage earner," and the trailer and other forms of low-cost, mass-produced housing seem to be part of that way of life.

This was the kind of dwelling I saw in the cities and towns of New Mexico, and in the industrial communities. I naturally associated it with urban working-class areas, so it came as something of a surprise to see those prefabricated houses and trailers in remote villages. What reconciled me to their presence, whatever their style or lack of style, was that they were being used, being lived in, in much the same manner as were the older houses, and when I went into a few of them I was entirely reassured. The resemblance between the lifestyle of the younger villagers and that of their families or grandfathers whom I had known a half-century earlier was striking. The houses were much more comfortable, much healthier than the old ones ever were, and they were better furnished. No wonder the families were proud of them and glad to show them off. I sensed that certain traditional relationships—between the house and the family, the house and the community, the house and the place of work—had changed little or not at all. They were much the same as they had been for generations in the old adobe houses. I was satisfied in that these brand-new houses or trailers were bona fide vernacular.

I doubt that many contemporary American students of the vernacular, or many architectural historians, would agree. As I mentioned earlier, the usual scholarly approach to the vernacular is to concentrate on the construction of the house, the materials and techniques used, and on the geographical or ethnic origin of its structural features. We have produced over the last generation impressive literature on what might be called architectural archaeology, and we have a much wider knowledge of house-types, often of a very obscure kind. But the emphasis has been on the old, the pretechnological structure. From the architectural point of view that is probably more stimulating to the student. What makes me uneasy is that the word *vernacular* now covers many lively aspects of popular culture, especially contemporary popular culture, and by concentrating almost exclusively on the anatomical aspects of old buildings, the field of vernacular architecture studies runs the risk of being antiquarian.

The solution, as I see it, is to explore the history of the vernacular dwelling, not merely here in the United States but as far back as we can go—at least in the history of the Western world. When we eventually undertake this, we will discover that the vernacular dwelling, the dwelling of the laborer or peasant as distinguished from the dwelling of the aristocracy, has been the subject of constant control and regulation, and at the same time the recipient of definite rights and privileges. Far from being a small and primitive version of the house of the nobleman or merchant, it has been a distinct form with its own rural way of life.

Americans are reluctant to discuss class distinctions in our culture, probably because we feel this would be undemocratic and adversarial. Nevertheless, we all recognize the difference between the houses lived in by the working class and those lived in by the rich. The difference is often less a matter of size and cost than of how space, interior as well as exterior, is organized and used. The average white-collar home is likely to contain a great variety of what anthropologists call monofunctional spaces: spaces of one kind or another set aside for a special use or a special person. This was always characteristic of the aristocratic household, even in the remote past, but its prevalence in the houses of the middle class is relatively new. Many scholars have discussed the development of the elaborate floor plan in domestic architecture, Philippe Ariès and Yi Fu Tuan among them. The nineteenth century seems to have been the time when the obsession with monofunctional spaces or rooms reached its climax. The contemporary middle-class dwelling manages to survive with fewer spaces but, in the guise of a free flow of space, new ones keep emerging: media entertainment centers, hobby rooms, exercise rooms, and super-bathrooms. The modern hi-tech kitchen is becoming a cluster of monofunctional spaces.

There are several things about this segmentation of domestic space that I find interesting. In the first place, it is merely a small-scale architectural version of a widespread modern tendency to organize *all* spaces in the landscape in terms of some special function. (It also recalls the tendency in prehistoric Pueblo architecture to define all architectural spaces in terms of content.)

But equally significant is the fact that the working-class house has been largely immune to the appeal of the monofunctional space. The house may well contain many rooms, but most of them serve several uses, uses which can change from hour to hour or from day to day. The garage serves as a storage room, then becomes a workshop. The kitchen is where we watch television and cook and eat; the dining room—if there is one—is for homework. The out-of-work brother-in-law sleeps on the living-room couch, and the men in the family tune up the second-hand car on the patch of lawn. These are strictly temporary expedients. All, or almost all, spaces in the house can be shared and used in a variety of ways. This reflects what I would call a vernacular concept of a space: a space has no inherent identity, it is simply defined by the way it is used. The middle-class or establishment concept is almost the direct opposite: each space is unique and can in fact affect the activity taking place within it. So, in the design of domestic spaces and their relationship, the skill of the architect and planner is always called for.

Nowhere is the contrast between the planned, specialized organization of domestic space in the establishment household and the fluid, undifferentiated spaces in the working-class dwelling more striking than in the provisions each of them makes for hospitality. When we enumerate the spaces and symbols devoted to hospitality—or exclusion—in even the average middle-class dwelling, we are likely to be surprised by their profusion: the formal front door with doorbell or chime, the formal lobby or entrance hall with clothes closet or so-called powder room, the drawing room (or library), the guest bedroom and bath, and formal dining room, to say nothing of parking spaces for visitors. Each of these spaces contains discrete symbols of family status; all of them together constitute a sizeable fraction of the area of the house. Along with this lavish use of space for hospitality often goes a very demanding domestic schedule of hospitality, fixed many days in advance. All of this organization for hospitality has its justification: the guest is ceremoniously introduced and admitted to a private and exclusive domain: to a territory of which the host or hostess is sovereign. We may criticize the formality of what we now call entertaining, but even its modest contemporary version represents a distinctly establishment definition of the house as an autonomous, self-sufficient territory, a focus of power and influence, a space where the stranger is formally admitted to be a member of a group.

The vernacular dwelling knows no such tradition. Its hospitality, though no less generous and welcoming, is informal and unpremeditated: no special rooms, no special days or hours, no special china or special cooking area called for, and the guests who appear, often uninvited, are not there for negotiating alliances or soliciting favors: they come to be included in the daily routine of the family.

I find nothing to criticize in this. It seems entirely consistent with the vernacular concept of the dwelling as a refuge from the workaday world, a place for the rituals of privacy, not for the pursuit of influence and power. As I have tried to indicate, the wage-earner dwelling delegates as many functions as it can to the public realm, reserving for itself the role of providing shelter and perpetuating family awareness. Unlike the middle-class house, the vernacular house is not a jealously guarded territory, and the outsider undergoes no entrance examination. As a member of the extended family or of the neighborhood, he or she is automatically included in the domestic order.

Hospitality, in short, is less an initiation into the house as an autonomous territory than it is a celebration of the super-family, and the best kind of celebration, the most generous kind of hospitality is that which is staged *outside* the home. The graduation party, the wedding reception, the grandparents' anniversary, the family reunion take over the school gymnasium, the parish hall, the hall of the local protective fraternal order, and for the time being the super-family uses it as if it

A car decorated for the wedding couple. (Author's photo)

belonged to them and no one else. From behind the closed doors come sounds of revelry: of flash photos, of laughter and long, emotional toasts. Benny Vigil and his Rock Caballeros play from eight in the evening until dawn. Outside in the darkness a shiny car, decorated with crepe paper flowers and streamers, waits for the bride and groom, and off they go for a weekend in Las Vegas.

This is the kind of event and the kind of space the vernacular dwelling *has* to have to survive. Its dependence on its immediate environment is not a servitude, it is something that can always be counted on, something morally dependable. For that is what distinguishes vernacular space from territorial space: it belongs to *us*. We have no legal title to it, but custom, unwritten law tells us we can use it in meeting our daily needs. Vernacular space is to be shared, not exploited or monopolized. It is never a source of wealth or power, it is in the literal sense of the term a common ground, a common place, a common denominator which makes each vernacular neighborhood a miniature common-wealth. Thus the contemporary way to study the vernacular dwelling is to see it not as an autonomous realm but as a structure which achieves completeness by relating to its environment.

A traditional festive use of the forest environment. (From *English Forests and Forest Trees*, London, 1883)

6 Beyond Wilderness

In the 1840s New York was a city of more than two hundred thousand, but it had managed to keep many of its ties with the surrounding landscape of rivers and beaches and forest. All day long, yawls and small sailing boats and yachts moved up and down the Hudson and across the bay to Long Island and the wooded shores of New Jersey. Flocks of ducks and geese and even swans hovered over the marshlands, and the sandy coast of Long Island from Montauk to Jamaica, deeply indented by bays and inlets, though with few inhabitants, was much visited by excursionists in search of trout and bluefish. Fire Island and Islip and even Barnegat in New Jersey were popular among holiday sailors. A few years earlier a ship loaded with wheat had sunk near the entrance of Egg Harbor and thousands of waterfowl floated on the water. When the day's fishing and shooting were over, families had clambakes and feasted on what they had caught. Often when the small boats sailed back to the city, they carried a cargo of fresh fish and fresh game—easy to sell in the markets and on the streets of Manhattan.

Long Island had many deer and only recently the last wolves on the northern end had been exterminated. But many city hunters preferred to cross the Hudson and explore the rugged country in Rockland and Hudson counties, where game was plentiful.

When in the 1830s the railroad arrived in New York and began to reach out to the north and west, there was already travel by steamboat up the Hudson as far as Albany—mostly pleasure travel to admire the forested mountains and the valley whose grandeur, so it was repeatedly said, rivaled that of the Rhine. The more prosperous and adventurous hunters were thus able to go into the Catskills and even as far as the Adirondacks. They usually found themselves in a frontier landscape where settlers were only beginning to clear sections of the dense forest, and where the profusion of deer and bear and wildcats and wolves was a problem. The measures the pioneers took to protect their crops and livestock were drastic: In the severe winter of 1836, thousands of deer were entrapped and immobilized by the deep snow throughout the inland Northeast; they were immediately bludgeoned to death by bands of club-wielding men; the bodies were so numerous that only the hides were collected. A few years later, with the advent of railroads, the backwoods population, still exterminating the local wildlife, found that it was possible to ship freshly killed game to the city, and there evolved throughout remote sections of the eastern seaboard a profitable form of commercial hunting: furnishing the city markets with game from the wilderness.

Many city residents were outraged by the slaughter, but in fact the destruction of wildlife and cutting down of trees had been part of pioneering ever since the first colonists set foot in America. Wolves were so numerous in early Massachusetts that Governor Winthrop found it prudent to carry a gun with him whenever he walked about Boston after dark, and his son was the first to import a pack of hunting dogs. In time the towns around Massachusetts Bay and in coastal Rhode Island, concerned for future supplies of game as a source of food, imposed closed season on the killing of deer and certain birds, and on the collecting of valuable fish. The newer, more isolated communities further inland resisted any such restrictions, protesting that they were still in danger from predators; they suspected that the laws were designed by rich city dwellers to suppress poor farmers who hunted (so they said) simply in order to feed their families and protect their livestock. To them, the closed season resembled the feudal laws against hunting in the royal forests in England.

Was that their only reason? Since the beginnings of man's existence on earth, the wilderness has engendered two sorts of reactions: hostility and fear on the one hand, a sense of protection on the other. Russell Lord, in *The Care of the Earth,* attributed the love we often feel for the forest "to racial memories beyond the farthest rim of present consciousness. The primates or great apes from whom it is now generally acknowledged we are ascendant were undoubtedly arboreal in habitat and frugivorous and vegetarian in diet." [1] But anthropologists, using much the same sort of prehistoric data, come up with an ingenious explanation of why we might *resent* trees in general. Their reasoning is this: After the retreat of the last Ice Age, fifty or more thousand years ago, the inhabitants of the Northern Hemisphere found themselves living for many centuries in a very brisk, not to say subarctic world, totally devoid of trees. It was a vast, open, wind-swept landscape of grassland where

Woodcut from "The Dream of Poliphilo" shows the dreamer fearfully entering a dark forest —perhaps representing entrance into the unknown. (From C. G. Jung, *Man and His Symbols*, New York, 1964)

giant bison, mammoths, reindeer, wild horses grazed, and where man learned to domesticate and eat certain grasslike plants, wheat and barley and rye. As the climate became warmer, trees started to grow: first birch and willow and aspen, then pine and oak and beech and other shade trees. Those once familiar herbivorous animals moved away, to be succeeded by others preferring the forest: deer and stag and boar, and bear and wildcats. The part-time farmer and stockman sought out grass-grown clearings in the forest where he could graze his livestock, plant his grassland crops, and make himself a house. The surrounding forest became an alien environment to be fought either by destroying it bit by bit, or by setting fire to it. "Man, the user of fire," Carl Sauer wrote, "from time immemorial has modified vegetation, made and maintained savannas, prairies, and other grassland."[2]

But ecology, when people make decisions, does not tell the whole story. When it comes to understanding and accepting a landscape for what it is, history, even fragmented and distorted history, is more reliable than theory; and recorded history in Europe makes it clear that beginning after the fall of Rome, the primeval forest was recognized by some as an effective political boundary, a defense against outside enemies and a way of keeping a community intact. When those small, grassy enclaves grew large enough in the Dark Ages to rank as fiefdoms, baronies, even counties, then they had to be protected, and what better way than by preserving the almost impenetrable tangle of trees, dead or alive, vines, marshlands, infested by wild animals, that surrounded it? When the forest acquired a military and political function, it automatically acquired a military and political status: castles, that is to say, walls and towers and trails; and a body of armed men, many on horseback, acting as guardians—and ultimately becoming the rulers. Those lowly farmers and stockmen with their routine of heavy work and their denuded fields and pastures and crude villages needed defense, notably in the matter of keeping wild animals away from crops and livestock: so the corps of mounted protectors became part-time hunters.

Charlemagne in the ninth century organized hunting expeditions, sometimes lasting for months, to rid the landscape of destructive wildlife; and the grateful peasantry served as beaters and grooms. Two centuries later, after the Norman Conquest, hunting became the ruling passion of the nobility; and that was when the word *forest* came into the English language—to designate not a kind of vegetation or topography, but a legal entity: an area outside (in Latin, *fores*) the realm of customary law; an area reserved for certain important persons, the king included, who enjoyed hunting big game. Commoners were subject to ferocious penalties if they presumed to hunt without permission or to poach.

Wilderness thus became the domain of the nobility, an environment where they alone could develop and display a number of aristocratic qualities. Friction arose between the peasants—inhabitants of open, unobstructed outdoor spaces—and the noble occupants of the forest, and that friction persisted as long as the peasant felt excluded from a portion of the landscape that he believed was his by right of heritage. Only recently have we learned that those cruel laws persisted in England well into the eighteenth century; indeed, some of the worst of them were

King Harold hunting, from the Bayeux tapestry. (From Michael Brander, *Hunting and Shooting,* New York, 1971)

drawn up and enforced, not in the Middle Ages, but in the Enlightenment: to keep the rank and file from entering the beautifully landscaped, carefully planted hunting parks at Windsor and other estates. The settlers in New England probably knew of those restrictions, and were human enough to avenge themselves, after a fashion, by destroying the forest wilderness in the New World.

And, after *their* own fashion, the prosperous amateur hunters in New York reacted against rural lawlessness. In 1844 a group of prominent New Yorkers, fearing that the unbridled slaughter of game in the back country eventually would mean no game for *anyone,* formed the first conservation group in America: the Association for the Protection of Game. It had two objectives: outlawing the public sale of game in the city markets, and encouraging amateur field and forest sports among all Americans.

Among the founding members of the association was a young Englishman by the name of Henry William Herbert. A member of a

Shooting prairie chickens in nineteenth-century Kansas. (From Brander, *Hunting and Shooting*)

distinguished titled family, schooled at Eton and Cambridge, he made his living teaching Latin and Greek in a small private school in a new and fashionable section of the city called Greenwich. He had arrived in America in 1831 hoping to become a writer, and was so dedicated that within four years he had written the first of several historical novels, translated several works of Dumas and Eugène Sue, produced a body of poetry which one critic believed to have "considerable merit," and started a literary magazine, the *American Monthly,* partly out of resentment when the *Knickerbocker* rejected a contribution of his. In 1839 his reputation was secure enough for him to contract for a series of essays on sports in England and America for a popular magazine, the *American Turf Register*.

Herbert chose to protect his reputation as a writer of serious literature by using the pseudonym Frank Forester for his ephemeral sporting essays. In England, where his subsequent novels sold many more copies than in America, he continued to be known as H. W. Herbert, the nephew of the earl of Carnarvon, and in fact he never gave up his English citizenship. But in this country he was less popular as a poet and classical scholar—he produced a verse translation of two plays of Aeschylus—than as Frank Forester, the genial and prolific writer of books on field and forest sports. All in all he produced seven such books, many with his own very competent illustrations, on fishing and hunting and horses and horsemanship. His last, a *Complete Manual for Young Sportsmen,* was still in print a century ago. During his lifetime, collections of his poems and short stories and essays were published in England.

Herbert seems to have been a difficult young man, unhappy in his relationships, often involved in lawsuits and quarrels, and always homesick for England. Those who knew him (and came to dislike him) described him as a poseur, always bragging about his aristocratic background, and critical of American democracy. He appeared on the street in outlandish sporting dress, with enormous spurs, in a checkered suit with a Scotch plaid shawl thrown over his shoulder. He let it be known that he belonged to a celebrated hunt in Yorkshire—"perhaps the most sporting county in England"—and that his father had owned two packs of hounds. He married twice; by his first wife he had a son who later went to England and served as chief of staff in World War I. He built a mock-Gothic house in northern New Jersey and brought his second wife there. She left him after three months. "Heartbroken," according to the notice in the English *Dictionary of National Biography,* "he invited his friends to a dinner [at a New York hotel]. Only one person came. After dining Herbert shot himself through the head." He was fifty-two years old.

He had two identities and two distinct literary careers, but his life had been dedicated to a single purpose: the celebration of an aristocratic country way of life such as he had known as a boy in Yorkshire and at Eton. It was a life based on the possession and exploitation of ancestral land, and on field sports, particularly on fox hunting. The medieval hunt had been military in its discipline, its hierarchical organization, and its rigid delegation of duties. It had made much of the relentless pursuit of large and dangerous animals, and its ritual killing, its final ceremonies were steeped in a macho mysticism of blood and death.

But fox hunting, as it evolved in the last decades of the eighteenth century, was at once more civilian and more civilized. It was socially exclusive and assumed instinctive compliance with unwritten rules of dress and deportment, but gentlemanly rather than military standards of courage and courtesy and restrained competitiveness were the order of the day. It is worth recalling that the sport became popular after the enclosure movement in England had produced a landscape of hedges and open fields with much less woodland, a landscape largely populated by prosperous, independent farmers who had their own hunts designed to eliminate foxes—considered an destructive form of vermin. In the early nineteenth century legislation required that every hunt have the permission of all landowners involved, and according to some chroniclers of the sport, the result was that many farmers welcomed the hunt and even took part. It is, however, hard to believe the same writers when they state that the foxes themselves (when they survived) were not unduly frightened and actually relished the chase.

These were the sporting traditions which Herbert had not only inherited but which he resolved to pass on to American sportsmen. It was only in the plantation South that elements of the English hunting tradition survived. Elsewhere in the East, in the pioneer back country, on the farm, and even among amateur hunters from the city, the only respectable hunting talent was marksmanship: American hunters were widely admired for being excellent marksmen. But this was to be expected: in a landscape where game was abundant, all that the amateur hunter had to do was to wait for a likely target—preferably an edible bird or beast—to appear, take aim, and then watch it fall. As a consequence of this casual attitude, the average American saw no need for any special training, any special equipment, or indeed any familiarity with the habits of wild animals. A further consequence was that he knew nothing of the excitement and challenge of hunting in the wilderness, or of the delights of knowing the forest in its seasonal variations. To Herbert and his English colleagues, the joy of prolonged contact with a familiar ancestral environment, of comradeship with other hunters, and not least the satisfaction of abiding by time-honored customs, were the most valuable elements in the hunt; and these were what he resolved to inculcate in his American followers.

Herbert was not a great writer, though fluent and correct; but he was a good teacher, and in his writing he knew how to insert all sorts of information and instruction in the midst of his vivid accounts of hunting in the Canadian wilderness or fishing in the Catskills: the name of the best English firm for rifles, the proper shirt to wear, and how a "collection of horses is a *stud.* The application of the latter term to the male horse is not merely *vulgar* squeamishness, but sheer nonsense." He told fishermen that wearing gloves while fishing was no less outrageous than wearing an overcoat at a fox hunt. For the benefit of would-be campers (they were few at that time), he included some twenty recipes for meals to be prepared over an open fire: among them halibut *à la Provençal* and haddock *à la Walter Scott.*[3]

Still more helpful were Herbert's pronouncements on the ethics of hunting. The animal being hunted must be *game,* not vermin like squirrels and rabbits and raccoons. The animal must be dealt with according to the laws of chivalrous and honorable sporting. The hunt must involve both skill and nature knowledge, and, last, there must be either peril or competition in the whole action.

One other reason for Frank Forester's popularity was his recognition of the aesthetic qualities of the wilderness environment. The old Puritan objection to the forest as a satanic realm, teeming with anti-social, anti-Christian temptations and pitfalls (never powerful outside of New England), had been pretty well forgotten by the mid-nineteenth century. The old resentment of feudal restrictions on wilderness exploitation no longer mattered; and though the frontier hunter was likely to prefer disposing of wild animals in a deforested environment, the urban middle-class amateur hunter—Frank Forester's target public—welcomed his emphasis on the less utilitarian benefits of the wilderness experience. "Nothing is lovelier in nature," Forester wrote in his discussion of deer hunting, "than the lone passes of the Adirondacks highlands, with all their pomp of many colored autumn woods, poised tier above tier into the pale clear skies of Indian summer with all their grandeur of rock-crowned peaks. . . . Here the true foot, the stout arm, the keen eye and the instructive prescience of the forester and mountaineer are needed; here it will be seen who is, and who is not, the true woodsman." And then he adds (as if to remind the amateur hunter of the essentially macho character of the experience), "the most rapturous of all, the moment when the quick rifle cracks and the stricken hart bounds aloft, death-wounded, and falls headlong."[4]

In those descriptive passages the reader is of course supposed to hear the voice of Frank Forester, the fearless hunter and wise counselor in matters of woodcraft; but anyone who knew about Henry William Herbert recognized at once that it was he who was giving an Americanized account of upper-class hunting in the English landscape. For the inexperienced American reader, Herbert mentions as normal the presence of many other people in the forest: the commercial hunter looking for fresh meat and hides to sell, the small farmer in the midst of the stumps of his new field; the backwoods logger, the squatter, the Indian: these substitute for the bystanders who are part of every English hunt, the small crowds of villagers and farmhands and neighbors who come to admire the gentry in their pink coats and their fine horses and hounds. Likewise, his often detailed descriptions of different aspects of the forest, at different seasons, different times of day, are his way of recalling the ancestral familiarity with the hunting country, evidence of *belonging* in the landscape. Herbert was by no means the only writer of his generation in America to undertake the education of the urban sportsman. Indeed, some of his accounts of frontier barbarities in hunting deer *very* closely resemble accounts of such events written by John James Audubon a good ten years earlier. But none of those other writers had Herbert's English perspective, and none wrote so well. That was probably why he (or Frank Forester) was so popular, so long remembered. Prosperous young Americans learned from him how to dress, how to talk, how to behave in the still unfamiliar world of hunting. And undoubtedly he set an example for American hunters of every class. They learned to reject the heavy-handed methods of their pioneer forebears. Even today, small-scale, small-game hunting is often more enjoyed for the forest experience, and the sense of conforming to a tradition, than for the bloodshed. It was Herbert (and to a lesser degree his writer colleagues) who helped integrate the wilderness into the everyday American landscape.

That is to say, Herbert played a role, a small one generally overlooked, in the development of an American romanticism. But for any school of romanticism to be popular it has to have a racial or ethnic ingredient. Everywhere and at all times humanity has been emotionally affected by the immensity and mystery of the forest. We have feared it as the abode of everything dangerous. We have worshipped it as the most perfect of God's creations. Either way we have gone on describing it, writing poems about it, and even venturing into it to find the truth. When America was a very young nation, we made friends with the forest: the threat of Indian raids was a thing of the past, and we forgot the Puritan warnings. Early in the century we therefore witnessed an effusion of forest poems, forest odes, forest painting, some of it very good. But it was only when we realized that the American forest was *different*—

not only in its extent, its immense variety, and its brilliant autumnal beauty—that we considered it a unique national treasure.

Thomas Cole, the founder of the Hudson River school of landscape painting, remarked that "the painter of American scenery has indeed privileges superior to any other. All nature here is new to art." He reminded his fellow artists that they had "a high and sacred mission to perform. . . . The axe of civilization is busy with our old forests and artisan ingenuity is fast sweeping away the relics of our national infancy." In 1830 a pupil of his, Asher Durand, undertook to publish a series of folios devoted to "picturesque views of American scenery," designed to teach the public that "the wild grandeur of our western forests . . . [is] unsurpassed by any of the boasted scenery of other countries."

Yes, but what made our forests uniquely American? Simon Schama suggests that the stereotype of the forest home as a benevolently presocial space protected from, or in active resistance to, urban, Roman concepts of civility, law, and the state, "was an ancient Germanic tradition, at home in the England of Robin Hood," and a cultural stowaway brought over to America.[5] When artists and poets praised the forest, they were implicitly praising the forest home of the pioneer, thoroughly isolated from the city. The irony was that establishing such a home inevitably meant the destruction of part of that same forest, the resumption of the age-old conflict between forest and open outdoor space. But the sanctity of the home was well worth the cost, and an essential characteristic of the American forest was its privatization, its fragmenting into countless small private holdings. As the destruction of the immense common forest proceeded, it became customary for every homestead to reserve a portion—usually a quarter—of the land for a woodlot, a source of wood, of course, for fuel and lumber, but also a place for private hunting—rabbits, woodchucks, squirrels, an occasional turkey or quail: vermin in Frank Forester's parlance. The woodlot nevertheless was a genuine miniature forest, visible on the farm landscape of America until about a half-century ago.

That fragmentation, noted even in early colonial times, came about because we lacked any memory of the communal forest, essential in any European landscape, and also of necessity lacked any common historical or mythical memory, holding us all together. Goethe, in distant Weimar, seemed to discern how fortunate we were in those lacks:

America, you're better off than
Our continent, the old. You have no castles which are fallen
No basalt to behold.
You're not disturbed within your inmost being
By useless remembering and unrewarding strife.
And when your children begin writing poetry
Let them guard well in all they do
Against knight-robber and ghost story.[6]

Both Irving and Cooper sought to give a legendary color to the American forest, the first by introducing Rip van Winkle and Ichabod Crane, and the second by introducing Nattibumpo and other noble Red Men; but the American public—that is to say, the prosperous traveling public—wanted no Gothic element in its romanticism. The forest stood for holidays and relaxation and an occasional contact with the sublime (as at Niagara Falls) or for a pedestrian adventure such as a climb up Mount Washington for the view. Those who traveled to see the splendors of the American forest, whether up the Hudson or into the remote South, rarely went for long or entirely alone. Thoreau, whom an earlier, less scientifically minded generation of environmentalists admired for his love of nature—and particularly for his cryptic statement that "in wilderness is the preservation of the world"—rarely ventured into the back country without a companion and, far from being distressed by evidence of human occupation, found it well worth examining—in the case of the unattractive Indians he came in contact with in Maine.

What has been said (with critical intent) of Thoreau might well be said of most of those pre–Civil War adventurers into the American wilderness. He refused "to ponder the significance of the whole assemblage of his perceptions and reactions to a subject. . . . He gathered together materials on a given subject—autumn leaves, let us say—or from a given sequence of experiences . . . and strung them together without attempting to synthesize them. . . . It seems to have been from a constitutional aversion to reconciling diverse experiences, particularly when they clashed with long and deeply held beliefs."[7] Translated into the more modest reactions of the average tourist exposed to the unfamiliar wilderness, Thoreau's reaction meant that no profound or permanent spiritual transformation was likely to take place; no sudden conversion to nature worship: the traveler returned from the adventure refreshed, exhilarated with much to tell, but as a human being no less committed to society. This polite but unmistakable refusal to go native, to "blend into the environment," this stubborn, old-fashioned humanism is one of the endearing characteristics of early American romanticism: best illustrated, perhaps, by the presence on a ledge in the Catskill forest overlooking the Hudson of a large and conspicuous (and successful) specimen of Greek Revival architecture, a hotel visible many miles away. The contrast between this monument to civilization and the monstrous birdhouses built by the Park Service to "be in keeping with the surrounding forest" typifies two very different approaches to the natural environment and our place in it.

Contemporary environmental or landscape historians have been ill advised in concentrating on the cult of the forest in nineteenth-century America, on the repeated accounts of public reaction to Niagara Falls and the Rocky Mountains. Except to deplore the wholesale destruction of trees in the pioneer midwestern states, they have little to say about the landscape produced by the creation of open spaces; nor do they give proper attention to the vast amount of tree planting that took place. If every mile of new railroad involved the cutting of many acres of trees, the railroad companies planted many acres to young trees, and according to the testimony of travelers every farmer, in addition to his woodlot, possessed orchards and ornamental planting in his yard.

The contradiction is only apparent: the most ardent and ruthless of tree bashers had a high regard for "useful" trees: trees that provided fruit, provided shade, indicated status, and which he could tend and control. It was in the many nurseries which came into existence in the first decades of the nineteenth century that arboriculture experimentation took place and that gardeners received something like a vocational education. Andrew Jackson Downing is a famous instance. If the study of nature means anything more than the study of the wilderness—and most current writing leaves one in doubt—it also means the study of crops and soils and weather, and the landscapes farmers and ranchers and gardeners create: it always includes the human contribution; and a great deal of practical and useful nature literature of that more inclusive description has always been produced.

When did that citizen-oriented romanticism go out of fashion? I suggest that it is still strong among Americans; though we are for the most part at home in the flatlands, in the open, organized environment of farms and offices and cities, we still revere the forest—perhaps more than any other people, with the exception of the Canadians—as a place for recreation and rest and rich impressions. The federal government, for all its shortcomings, instinctively understood the popular approach to the forest when it established national parks, beginning with Yellowstone in 1872—"as a public or pleasuring ground for the benefit and enjoyment of the people"; not for exploitation, not for the people's education in botany or ecology, but for pleasure.

The national forests followed suit. The full impact of those decisions was not felt until the early twentieth century, when visitors to the parks and forests started to come by automobile; and that was when the latent feudal impulse—to consider every forest the exclusive territory of the hunter and rich landowner—produced a number of organizations dedicated to defending the wilderness against the hordes of city holiday makers. This was to be done by limiting visitor facilities—notably roads and campgrounds—or by defining the forest as a special game preserve. The Boone and Crockett Club, the rich and prestigious successor to Herbert's Association for the Protection of Game, gravely suggested that the national parks should of course be preserved in their original prehuman condition so that scientists could conduct "Darwinist" research on heredity and environment. The membership of the Boone and Crockett Club included many ardent believers in racial purity, among them Madison Grant, and consistent with that belief went on record as strongly opposing the "introduction of non-native plants and animals into the National Parks."[8] In its earlier years, the Sierra Club had racial quotas in its membership.

The Sierra Club, although not the largest of conservation organizations, is the oldest and probably the best known. It was founded in 1891 by a group of well-heeled San Francisco nature lovers, admirers of the eminent California nature writer John Muir. Its original stated purpose was the study and protection of national scenic resources, wilderness areas, forests, and streams. At first it was an informal organization of men and women fond of hiking, exploring the Sierras, taking landscape photographs, and attending lectures by fellow members. But in the 1950s, because of a change in leadership, it became a dynamic, very vocal organization, largely devoted to defining and promoting a "wilderness ethic."

In government parlance a wilderness is a large tract of land left in its original uninhabited condition. In that sense there was and still is plenty of wilderness in the United States. But the Sierra Club preferred a more inspiring definition: a "region which contains no permanent inhabitants, possesses no possibility of conveyance by mechanical means, and is sufficiently spacious that a person crossing it must have the experience of sleeping out." In time such wildernesses as fell within the national parks or forests were set aside from areas meant for popular recreation and were called "primitive areas." Since the Sierra Club strongly disapproved of what the Park Service was doing to accommodate vacationers and tourists, it chose to confine its attention to the primitive areas, and fought to have them extensive enough to assure the wilderness visitor almost total isolation from the outside world and even from the more mundane visitors. The most revealing study of the Sierra Club and its philosophy, and of its emphasis on the wilderness experience, is a book by Linda Graber, *Wilderness as Sacred Space* (Washington, D.C.: Association of American Geographers, 1976).

Many environmentalists, particularly those called radical ecologists, set great store by this experience: to them it is like an initiation into religious brotherhood. To describe it briefly, the experience is a sudden or unforeseen awareness of the power of nature, and of our true relationship to it. "In wilderness," said Kenneth Brower, son of the former head of the Sierra Club, "we can see where we have come from, where we are going, how far we've gone. In wilderness is the only unsullied earth sample of the forces generally at work in the universe."

Strictly speaking, the wilderness experience is not mystical, for the relationship revealed is not with the divine but with unspoiled nature; but its effect is to subdue the omnipresent clamor of the ego and to reveal to us that we, along with all living things, are inseparable parts of the cosmic order. For all its intensity it is a temporary thing; we return to the everyday world, though spiritually transformed.

The trouble with this experience (as the example of Edward Abbey, once a popular environmental guru, shows) is that it all too often inspires the comparison between the primordial natural world and the world most of us live in. When confronted by the immense desert environment of the Southwest, Abbey, like everyone else, is rendered speechless—*bewildered* in the true sense of the word. He writes eloquently and earnestly, but sooner or later ego takes over, and the mountain eminence where he stands becomes a pulpit, and with renewed passion he denounces commercialism, desert desecration, urbanism, engineering, the consumer society, and overpopulation; on his return to the workaday world he continues, like one inspired, to bash civilization, history, and humanity. "I would rather kill a man than a snake" is his repeated boast.

The Sierra Club (and other environmental organizations) not only endorse these notions but publish them, and there has emerged a body of anti-urban, antitechnological, antipeople, antihistory books and pamphlets, all anthrophobic, all urging us to worship nature. "This love [of nature], when it sets up as a religion," C. S. Lewis wrote, "is beginning to be a god, therefore to be a demon. . . . If you take nature as a teacher she will teach you exactly the lessons you had already decided to learn."[9] In the case of the Sierra Club and of many well-intentioned environmentalists, nature is in fact teaching us that a return to origins, to the pretechnological purity of the past, to a static social order, is the only way to go.

Vociferous and threatening though those voices may be, it is well to remember that they represent a very small minority. After all, we are dealing here with a fragment of the environment not dedicated to work and the tasks of everyday existence but to pleasure and the satisfaction of emotional needs, and even then the sacred space of the wilderness is only one of many. Though it may seem as if the national parks and forests were overwhelmingly used (and abused) by flatlanders from the city, the state and county and municipal "pleasuring grounds," though only half as large in the aggregate as the federal parks, serve twice the number of visitors.

Since these smaller areas are located primarily to serve a weekend or holiday urban population, they are not always of great scenic or environmental value. In fact, many of them are entirely artificial. They start with a body of water (usually impounded), an area of primeval forest or second growth, and sometimes a natural landmark: balancing rock, waterfall, hill with panoramic view. Their attractions are also manmade: an agreeable and flexible composition of parking lots, playgrounds, trailer park, golf course, beach, and a sizeable area of natural forest and even renewed wilderness. There is likely to be an orientation center with a five-minute movie about the local environment, informative literature, lectures, and a rustic trail to the hilltop. The central visiting area is full of people of all ages. They lie on the grass, sleep, get tanned, practice aerobic exercises: they swim, they jog, they take brisk walks; they waterski and wander about to find or make friends. They pick flowers and take photographs. There is the constant sound of radios, the constant smell

(Photo: © Bruce Davidson/Magnum Photos)

of barbecue, a constant coming and going of cars and trailers. The park rangers and attendants wearily check to see that fires are put out, lost children found, trash put into containers, televisions silenced after nine in the evening, and among themselves refer in unflattering terms to the public and its predictable misbehavior. Wilderness buffs, passing through the park in their search for "meaningful" experiences, scornfully dismiss the activities of the crowd as mere recreation.

Recreation is indeed the right word: recreation as pleasure and relaxation, but also as a recharging of exhausted bodies and minds: recreation by means of a temporary contact with nature.

A favorite study among anthropologists and geographers and academic environmentalists is the manner in which various societies or historical periods "perceive the environment," which is another way of saying how they define the man-nature relationship. They study how the contemporary blue-collar urban American "perceives" his or her environment, and such a study often reveals some disconcerting characteristics: among them the fact that (because of racial and ethnic infusions from other parts of the world) the old northwestern European WASP concept of nature and landscape is fast losing its significance with the majority of urban Americans. Untouched nature is no longer seen as sacred; and it is no longer entirely green. Nature, instead, serves as a source of energy, almost always invisible and impalpable, but so powerful that direct contact with it can be dangerous. In consequence, it has to be filtered, diluted, made to conform to federal standards of health and safety. Sun and air and water, taken straight, can harm us; hence the "treatment" of water in pools, the sun lotions, the filtered "conditioned" air in our houses. Nevertheless, it is nature which provides us with health and vigor and peace of mind, and all of those activities we see in the recreation area are inspired by an urge to absorb, assimilate, and possess the invisible healing powers of the environment: its green resiliency, its constant variety, its reassuring plenty.

This vernacular, urban, contemporary perception of nature is a complicated matter; all that now seems clear is that we are formulating a demystified, demythologized definition which automatically includes human participation—if only as a moderating, rationalizing force. Just as the untamed wilderness is rejected as too unpredictable in favor of the grass-grown open space interspersed with trees and bodies of placid water, the anarchic sense of individual uniqueness is tempered among most of us by an awareness of the rights of others. A philosophy which exalts health and safety to the status of virtues cannot be called heroic, and religion seems to play a very small role in our definition of the purpose of life. But that may well come later.

In the meantime, this new synthetic version of the forest as the setting for human interaction with nature is evolving throughout the country. And with it come new forms of social interaction based not on competition and an exaggerated sense of territory or the sanctity of origins, but on the sharing of limited, temporary resources for health and well-being. As our population continues to grow, and as it becomes more non-European, we will turn more and more to these open, grass-grown public areas for contact with nature. The humanized landscape, much simplified, much impoverished, but still vigorous, will reappear in time, and wilderness will retreat into the background, to be still available to those who want a transcendental experience.

It would be unreasonable to hope that environmentalists and the environmental movement will overcome their fixation on wilderness in the near future. Too many spellbinders, long before Thoreau, have been at work persuading well-meaning lovers of nature that there can be no better goal than blending with the green environment. But the majority of Americans are unconsciously seeking a new definition of wilderness; part fact, part metaphor, much as the children of Israel and later the Puritan settlers feared wilderness both because of the temptations it offered to withdraw from civilization, and because of its perils. The experience of wilderness was of incomparable value: Judaic law and religion would be unthinkable without those forty years of wandering in search of towns and of water, and of the promised land. The trials of the Puritans lasted longer, and from the forest wilderness they eventually wrested a prosperous landscape and the vision of a just social order that still inspires us. But the wilderness experience is always an interlude, a moment of new insights. It is time that it came to an end, time that we undertook the reconstruction of our desolate cities and the reinvigoration of our rural communities.

An arbor fantasy. (From Robert Mallet, *Jardins et paradis,* Paris, 1959)

7 In Favor of Trees

Like millions of other Americans I have no great liking for wilderness and forest, but like the majority of Americans I am fond of trees: individual trees, trees in rows along the street or in orchards, trees in parks. I continue to plant them when and where I can—to such an extent that when their leaves start to fall I look forward to many months of raking and transplanting in preparation for the spring.

The value of trees is not only that they can be beautiful and that they give us shade and privacy and coolness in the summer; they also demand our attention and care. We are constantly interacting with trees: some of them give us fruit, others give us firewood, and all have to be thought about and even worried about when we consider the future. In brief, trees give us a sense of responsibility and sometimes a kind of parental pride; each domesticated tree calls for an individual response, a response far richer, far more rewarding than a strictly passive—aesthetic or ecological—response to the forest.

What geographers call the Atlantic landscape stretches across northwestern Europe—England, France, the Lowlands, Germany, and Scandinavia; and in the course of the last three centuries it has been transplanted to Canada and the United States. It can be thought of as the gradual creation of those Indo-European migrants who came out of Asia some seven thousand years ago with their livestock and who eventually occupied all of Europe. In addition to the Atlantic landscape north of the Alps, they also produced the Mediterranean landscape—equally varied and beautiful, but adjusted to a mountainous terrain, hot dry summers, and no great abundance of moisture. By contrast, the Atlantic landscape—both in America and in the Old World—is characterized by a green, rolling topography with many rivers and plenty of rainfall. Mexico has a version of the Mediterranean countryside, and so have parts of New Mexico and California.

Century after century the early Indo-Europeans wandered from the Ukraine to Greece and Norway and even Ireland. When they occasionally settled down, their livestock grazed in the surrounding forests and grasslands, and families raised small crops of wheat or rye or barley. They brought with them out of Asia certain fruit trees. Alma-Ata, the capital of Kazakhstan, means "father of apples," for the mountains in that part of Central Asia once contained immense forests of nothing but apple trees, as well as forests of pear trees and apricot trees. Those fruit trees, as well as certain nut trees, were greatly prized by the migrants, for they provided sugar and oil, as well as calories, and when planted (or transplanted) they symbolized the permanent home and family.

The forest was, of course, the dominant element in that prehistoric landscape—even the landscape of what we used to call the Dark Ages: the period between the fall of Rome in the fifth century A.D. and the Norman Conquest of England six hundred years later. It was a frightening and inhospitable place, extending from Poland on the east all the way (with numerous breaks, to be sure) to Holland—which, paradoxically enough, means "land of woods." We hear much about the density and extent of the Amazonian rain forest, but one of the largest and most impenetrable forests in the world is in northern Russia, of which it is said no one knows what fear is who has not been within its dark and tangled precincts. Nowhere was the early forest looked upon with anything but awe. Legend depicted it as the habitat of giants and elves and mythical creatures, a refuge for outlaws and dangerous spirits; and some of that legend persists in familiar fairy tales.

Nevertheless, the forest played an important role in peasant economy. It provided firewood, wood for building, and a variety of herbs and wild fruits; cattle from the village grazed in the grass-grown clearings, and under the many oak trees herds of pigs ate acorns and mast. Cultivation in the forest was forbidden, and the small-stockman, the small-farmer was largely dependent on the garden, the common grazing land, and on the trees planted by the family or the village.

One of the attractive features of the Atlantic landscape in the Middle Ages was its popular culture based on wood: the planting and care of trees which produced fruit or provided material for a great number of crafts. The forest still contained much oak, the most prestigious of trees; but the open landscape, the landscape of fields and meadows and houses and gardens, contained an increasing variety of fruit trees imported from elsewhere; trees whose wood could be used for household needs and for wagons and plows; trees which, because of their everyday importance, eventually acquired symbolic value.

(From Gaston Roupnel, *Histoire de la France rurale,* volume 2, Paris, 1973)

In his book *Sylva,* the seventeenth-century Englishman John Evelyn mentions an old German law which stipulated that "a young farmer must produce a certificate of his having set a number of walnut trees before he have leave to marry." Since the smelting of iron threatened in the seventeenth century to destroy many English forests, Evelyn suggested that what he called "iron mills" be established in New England. " 'Twere far better," he wrote," to purchase all of our iron out of America than to exhaust our woods here at home." [1]

Sylva, originally in four volumes, is an encyclopedia of arboriculture as practiced in pre-industrial Europe. Evelyn tells how to collect and plant the seed of numerous "useful" trees, when and where to plant, how to protect, trim, and feed them, and how finally to cut them and process the wood. Though the oak remained the king of trees—Evelyn devotes twenty pages to describing and praising it—he has much to say about each of the others:

> Elm is a timber of most singular use, especially where it may be continually dry, or wet, in extremes; therefore proper for water works, mills, the ladles and soles of the wheel, pipes, pumps, aqueducts, ship

> planks below the water line . . . also for wheelwrights, handles for the single handsaw, rails and gates. Elm is not so apt to rive [split] . . . and is used for chopping blocks, blocks for the hat maker, trunks and boxes to be covered with leather; coffins and dressers and shovelboard tables of great length; also for the carver and those curious workers of fruitages, foliage, shields, statues and most of the ornaments appertaining to the orders of architecture. . . . And finally (which I must not omit) the use of the very leaves of this tree, especially of the female, is not to be despised . . . for they will prove a great relief to cattle in the winter and scorching summers when hay and fodder is dear. . . . The green leaf of the elms contused [crushed] heals a green wound or cut, and boiled with the bark, consolidates fractured bones.[2]

In reading Evelyn, we discover two kinds of pleasure. One comes from reading a wonderfully idiomatic English, clear and unaffected, emphasizing the visible, tangible everyday aspects of his topic. The second comes from many glimpses of a vernacular, country way of life based on the skillful exploitation of a local resource: the growing and cultivation and processing of trees used for making houses and furniture, in home remedies, in food and liquor and cooking; trees planted in farm gardens, in orchards, along country roads, in clusters to provide shade for cattle; trees in stately double rows to mark the avenue leading to a noble mansion, or to a town; each with its own traditional value to the craftsman, the artist, the housewife, the builder. Many retained from the remote past a powerful symbolism. The linden or lime tree was the tree of justice, local courts being held in its shade; the yew and the cypress symbolized immortality, and the apple tree stood for domesticity: so much so that a French geographer suggests that the apple tree is the prime symbol of the Atlantic landscape, just as the olive tree symbolizes the landscape of Mediterranean Europe.[3]

Evelyn wrote his book because of his concern over the increasing scarcity of certain essential types of wood. The king's navy and the growth of transatlantic commerce threatened the English supply of oak. Foundries and mills and furnaces consumed more and more forest wood, and the needs of a growing urban population for fuel, as well as the destruction of many forests in wartime, all threatened the existing stands. *Sylva* was accordingly addressed to the great landowners of England, urging them to plant as many trees as possible in a wholesale manner. It should be noted that Evelyn nowhere recommends the traditional forest of mixed woods, the hunters' forest. What he advised, not only for England but also for the Continent, was the creation of systematic, commercial forestry; and that was the type of forestry which evolved in the eighteenth century.

It was in the English colonies in North America that the old vernacular culture of trees was given a fresh lease on life. The early settlers lost little time before destroying (for commercial use as well as for clearing land for farming) immense areas of virgin forest. That, in fact, is what had happened to the forests of medieval Europe, but in America—again as in Europe—the planting and cultivation of trees flourished as never before. For that is a distinction we must always make: the forest as a massive collection of trees of all varieties is seen as a resource, not as an environment. Whereas the single or planted tree is seen by most of us as a permanent, carefully tended element of the human landscape, valued as an object both of beauty and of sustainable exploitation.

In any case, colonial America found several new ways of using trees. We developed an improved type of ax, the water-powered sawmill, and learned to build houses and bridges and dams and even roads entirely out of wood. Following the example of the native Indians, colonists extracted sugar from maple trees. Early in colonial history, we undertook to plant trees along our streets and roads, for shade and shelter, and when independence came, many towns and villages celebrated the event by erecting liberty trees. When the Midwest was settled in the early nineteenth century, immigrant handbooks and other periodicals advised the settlers to plant orchards first of all, even postponing the planting of a vegetable garden.

We soon had plenty of food on the market, and plenty of wood was still available in the forests. So the culture of trees in America took a new turn: trees were planted chiefly for their beauty and symbolism. Starting in New England in the 1850s, where women's organizations were dedicated to the beautifying of towns and cities, a national enthusiasm for ornamental trees everywhere transformed the village square, the college campus, many country roads and graveyards. The landscaped cemetery composed of winding roads, groups of trees, and expanses of lawn was in

Planting trees in the Kansas prairie. (From USDA Yearbook, 1949)

a sense the reconstruction of the old pre-industrial landscape of legend. On the treeless prairie the farmhouse was surrounded by a grove of trees, and their bright autumn colors gave certain trees an almost symbolic value, unique to America.

This widespread cult of ornamental trees brought about an immense increase in the number of nurseries and tree farms. At the same time, the nation as a whole became increasingly aware of deforestation in many regions. Beginning more than sixty years ago, at the time of the Great Depression, state and federal government agencies launched vast programs of tree planting. Hundreds of thousands of saplings were planted to check erosion, to break the force of the wind, to provide habitats for wildlife, to control flood waters, to modify the climate. Millions of trees, whole forests, were planted for ecological reasons.

Two generations ago the word *ecology* was rarely heard, and to most Americans the very notion that forests—natural or artificial—could serve other than human needs was a revelation. I am old enough to have lived through those first large-scale ecological experiments, and in retrospect I think we generally approved of them—though there still lingers among Americans the ancient belief that the forest is there for us to exploit in the meeting of daily needs: for fuel, for food, for grazing, for hunting, and for escape from social restrictions. The national park or forest is still thought of in terms of recreation and camping, and to be reminded of the many ecological benefits of the forest simply confirmed the reality of that prehistoric prototype. As a result, those numerous planted groves and belts and forests were quickly assimilated into the landscape and their recent origin forgotten. In fact, many of the windbreaks have been destroyed—or harvested—by farmers totally unaware of their original purpose.

It could be said that the reforestation or tree-planting programs of the Depression years helped inaugurate the environmental movement in this country. In that sense they were part of a worldwide shift in attitude toward the natural environment. Here for the first time on an extensive scale, the landscape, or part of it, was being deliberately altered not to serve immediate human needs but to preserve the natural order. It is quite true that in the course of planting and reforestation many highways were landscaped, many rest areas and recreational facilities came into being. It is also true that our national parks, even when overrun with visitors, try to make us feel that as citizens we are inspecting one sample of our national estate. Nevertheless, our national and state parks actually provide us with only the faintest reminders of our earlier forest or wilderness experience.

The contemporary forest experience emphasizes the *visual* aspect, the scenic, the ecological, the photogenic. We are not to touch, much less pick up and carry away, any object we find of interest. We are tactfully told that we are not at home but in a museum; a museum,

(From *Rural Affairs,* 1869)

moreover, which is increasingly concentrated on ecological or geological or botanical phenomena. The risk of vandalism and destruction helps justify this hands-off policy, though the influence of current environmentalist policy—the determination to preserve nature totally undisturbed by man—has had its effect. For the fact of the matter is, humanity's closest and most productive relationship with nature derives from personal, physical contact, and from a desire to appropriate whatever attracts us. "Leave nothing behind, not even footprints," the environmentalists advise those of us who go into the wilderness. "Take nothing except photographs." The visual experience, the spectator experience, is the only one permitted.

Our true feeling for trees derives from an ancient source—from centuries of domesticating, improving, protecting, and loving those other forms of life which are part of our daily existence. Looking back over more than half a century, I am struck by our growing desire for trees in our domestic environment, by our desire to plant trees, regardless of their economic value, in order to express a variety of basic emotions: the need to celebrate the home, the need for beauty, the need for some living thing to protect and transform, the need to pass on to the future some sign of our existence. Ecologists encourage us in this enthusiasm, assuring us that the tree we plant will help cleanse the atmosphere, moderate the climate, and close the gap in the ozone layer. But John Evelyn, nearly 350 years ago, provided us with a better justification: "Men seldom plant trees until they begin to be wise; that is, till they grow old and find by experience the prudence and necessity of it. . . . 'Tis observed that such planters are often blessed with health and old age." He added, in a passage I take very much to heart, "I am writing as an octogenarian, and shall, if God protract my years, and continued health, be continually planting till it shall please him to transplant me to those glorious regions above, the celestial paradise—for such is the tree of life, which those who do his commandments have right to."[4]

(From Heidi Lehmann, *Volksbrauch im Jahreslauf*, Munich, 1964)

8 The Past and Future Park

(Photo: Barrie B. Greenbie)

In every city, in every town, even in every village in America I go to I expect to find an outdoor recreation area or what is usually called a park; and I am seldom disappointed. No matter how new and unfinished a town may be, or however old and poor, I know that it will contain, wedged in among the crowded blocks of buildings, a rectangular space with grass and trees and meandering paths, and perhaps a bandstand or a flagpole. Some of these spaces, like Central Park in New York or Golden Gate Park in San Francisco, are so famous or so beautiful that I would go out of my way to see them, but the average small-town park has a different kind of appeal. All the basic elements are there—the trees, the grass, the pathways—but it is so subdued, so naked, that I wonder who uses it, and why. It seems to be an archetype: a specimen of the original, timeless *Ur*-park as it might have appeared in the Bronze Age, or even earlier.

The truth is that the urban park is a newcomer to the landscape. In terms of greenery and design, no such space existed in ancient Greece or in Rome, and it first appeared in European towns not much earlier than three centuries ago. Medieval towns had, of course, a number of public spaces which in accordance with vernacular practice could be used for a succession of functions: the marketplace, the *parvis* in front of the church, the graveyard next to it, the place for processions and for executions. But no open space in the town was ever set aside—let alone designed—for such a vague purpose as recreation.

Rich and powerful families had their own orchards and gardens, hidden behind walls, stocked with rare plants from Asia or America, and adorned with statues, specimens of topiary art, and fountains which played tricks on unwary visitors. Expert gardeners were called in to make a formal composition of flower beds and walks and flights of steps, and the garden was often the scene of elaborate social events. In time, the owners were persuaded to allow certain reliable elements in the town to come and admire the gardens. They consented to this not simply

Public recreation in sixteenth-century Holland. (From Lehmann, *Volksbrauch im Jahreslauf*)

out of vanity—for a garden showed how rich a man was—but also out of philanthropy. They believed that the working class could be morally improved by being exposed to beauty and order and a display of good manners.

The designers of the gardens suggested that one way to educate the common people was to make broad, straight avenues leading out of the gardens, flanked by uniform trees with statues of famous men and occasional Latin inscriptions. The straightness of the avenues had the advantage of making them easy to police, and of offering an impressive view of the palace or castle or mansion. As a place where all citizens could spend their leisure walking to and fro beneath the trees, the *allée* at once became popular. At certain afternoon hours, the world of fashion made its appearance in carriages or on horseback, and all elements benefited from access to what was soon to be called the park. It was there that garden came together with city, greenery with architecture and social forms. Indeed, the allée in a more urbanized guise has survived to this day in many cities: Pall Mall in London, the Cours la Reine in Paris, and the Unter den Linden in Berlin are familiar examples.

The formal garden itself, the museumlike collection of curiosities, did not fare so well, and in the early eighteenth century it began to suffer from neglect. No doubt the rise of a prosperous and educated middle class, particularly in England, accounted for the change. The geometrical garden, identified with the French aristocracy, failed to offer privacy and intimacy and a chance for the amateur gardener to try his hand. Writers like Alexander Pope and Joseph Addison ridiculed the prevalence of straight rows of identical plants and the absurdity of shaping box to look like lions or dragons. Travelers returning from Italy told of picturesque ruins and the pleasing disorder in the Italian landscape. From paintings they had learned that the landscape was more than a collection of isolated structures and spaces; that it was a composition which even included light and shadow and the remoter background of sky. In its day this reaction was very novel. It marked an awareness of the wider environment, of nature in all its forms and colors. Those who stayed at home in England found pleasure in discovering winding lanes bordered by hedges and wildflowers, the unplanned grouping of trees in the countryside, and many among the more prosperous English began to think of owning a place in the country. Merchants and lawyers bought up small farms and woodlands, and even whole villages, with the idea of making money in specialized or commercial farming. At the same time they had visions of creating their own personal landscape, generous in scale, adorned with lakes and groves of artfully planted trees and distant views; they, too, had seen paintings. A feature of such an estate would be that, despite its openness and simplicity, it was entirely private and removed from the local workaday world: an oasis of rustic beauty.

Out of these various impulses and memories and aspirations there evolved in the middle decades of the eighteenth century a new kind of garden or park, sometimes called the picturesque landscape garden or the romantic garden. Though originating in England, it became fashionable in France and Germany and, eventually, in nineteenth-century

America. In the beginning it, too, had its absurdities: ruins and Chinese pagodas and dark caves where some venerable hermit might have lived. But it soon evolved its own standards of correctness and good taste, and the older classical garden was either forgotten or transformed in accordance with the new style. Generally speaking, it represents the style of landscape design that most Westerners today instinctively accept. It belongs in countries that have a moist climate, good soil, varied topography, and a belief in the value of privacy and outdoor living. Whether it was adapted to great cities and to urban poverty is a question we are beginning to ask.

Nevertheless, the romantic style of landscape has always had a good press. It has been approved by art historians and by the general public, and for good reason: it has taught us how to design magnificent outdoor spaces as well as the smallest family garden; it has taught us how to achieve beauty with the simplest and commonest of means: the landforms and vegetation everywhere surrounding us; and it has brought us closer to nature and made us more at home in wilderness and mountains. I am one of those who believe that our current guilt-ridden worship of the environment is a sign of moral and cultural disarray, but the romantic school of design is not entirely to blame for our exaggerations, and scholars are correct in saying that it has had an immense influence not only on what we now call landscape architecture, but on architecture and urban planning as well.

Critics often dwell on the *production* aspect of art—on the artist and his or her skill, and what the artist tried to convey; but a garden is more than a work of art. It has a utilitarian value, and how the consumer uses it has to be included in the critical verdict. I think it can be said that the privately owned picturesque landscape was almost perfectly adjusted to the needs of a certain fortunate class of consumers. It provided solitude, beauty, the uninterrupted, essentially passive, spectator experience of nature, a sense of ownership, and, above all, the ability to exclude, tactfully but effectively, all outsiders and all contact with the urban world. How the sensitive visitor reacted to this landscape was the theme of much polite eighteenth-century literature: he is alone or accompanied by his beloved, and wanders without apparent destination in search of the several emotional treats that the garden provides: melancholy, awe, terror, and sense of oneness with nature.

The all-important eighteenth-century allée. (From H. Boerhaave, *Index plantarum,* 1710)

The nineteenth-century version of the romantic picturesque landscape park. (From Mallet, *Jardins et paradis*)

Unlike the Baroque garden with its well-defined axis leading to the palace, or the religious sanctuary focused on the shrine, the picturesque garden had no dominant feature to draw people together: usually no more than an oval lawn. In this respect architecture failed to play one of its traditional roles, for the owner's residence sought to blend with the natural setting when it was not located in a secluded domestic area of the garden. "The effect of introducing buildings amongst artificially established rocks and cascades as part of the landscape," Peter Collins writes, "was . . . merely the first step towards establishing the general idea that rural architecture ought essentially to be thought of as making natural scenery more 'picturesque,' i.e. more like a landscape painting."[1] Architects adopted Gothic architecture (as was the case with A. J. Downing and his landscaped gardens), or (in the case of the *hameau* of Marie Antoinette) the rustic appearance of peasant farmhouses. As Collins observed, the gradual abandonment of architectural symmetry in the romantic garden had the effect of producing a more flexible interior layout and of making the country house, whatever its scale or size, more habitable and less conspicuous. In the late eighteenth century, the country house or villa, thanks largely to its landscaped setting, developed into a distinct house-type, while the building with a public or official status, insofar as its presence suggested the intrusion of the city, was discouraged. In one German romantic garden of the period, a temple to Mercury, proposed as a suitably classical touch, was rejected: Mercury was the god of commerce.

Do we need to be reminded that the picturesque garden culture of two centuries ago flourishes in many of our suburbs and resorts? It has been refined and simplified: the man of the family now goes to work in the city, and the once common experience of the sublime has been replaced by gloomy thoughts about pollution. But they come to much the same thing: nature is all-important in human affairs.

It is more difficult to explain the great increase in the extent of the picturesque landscape in precisely the environment it once feared and avoided: the city. By and large, almost all of our parkways and larger city parks are modernized versions of the romantic park, characterized by winding paths, varied landforms, pastoral lawns and lakes and groves of trees, and isolation from the urban setting. Many of the finest specimens of the style are to be found in the heart of cities in America and Europe.

The story of how the romantic garden or park came to be chosen as the official establishment public park has been neglected by historians of landscape architecture—perhaps because they are unaware that there was, and still is, a vernacular tradition in the design of places for predominantly working-class recreation. In seventeenth-century England the public had access to many aristocratic gardens and, though used in farming, many country paths and lanes leading through beautiful scenery were open to all.

The practice seems to have been common throughout prerevolutionary Europe: in Germany a number of towns were known for their attractive and well-maintained country walks, each with a recreational destination. *Volksgartens* were created near Vienna and Berlin and other cities. Though they still had a geometrical layout with radiating allées, they included a lively assortment of popular attractions: tents and booths where refreshments could be bought, merry-go-rounds, shooting galleries, bowling alleys, and dance music played by a military band. In 1836 Charles Dickens wrote a vivid account of how working-class Londoners spent their Sundays: setting out at an early hour and walking to where they found a coach or a steamer to take them to such places as Greenwich and Richmond. There they picnicked, walked through the nearby fields, played cricket, and celebrated in the many taverns before going home.[2] Paris had its *guinguettes,* or taverns with gardens, where there was wine and music, and in pre–Central Park days (and undoubtedly later) New Yorkers frequented vernacular resorts in Hoboken and Long Island and along the beaches. Reading between the lines of descriptions in newspapers and magazines, we can deduce that these resorts were rarely visited by more fastidious citizens. They were crowded and boisterous and sometimes violent.

Nevertheless, they possessed a valuable quality largely absent from the carefully designed "natural" spaces of the city park which eventually took their place. In a workaday, often farming environment, they offered the spectacle of interaction between work and play, between the private and the public realm, between producer and consumer, between urban and rural ways of life. The vernacular resort, on one day of the week, was a kind of marketplace for the exchange of ideas and information, a reincarnation of the traditional street which was gradually disappearing under the impact of the industrial city. In a sense those extramural places of recreation were the last remnants of the old vernacular culture: they had their own kind of music, their own kind of sports and games, their own kind of food, and their own code of manners.

The last decades of the eighteenth century saw the emergence in city after city of the picturesque park, open to the public. Each park came to represent the impact of three distinct social forces: the urge to improve the living conditions of factory workers in crowded industrial centers; the urge to bring all classes in close contact with the moral and physical benefits of a "natural" environment: and the urge to improve the real estate values of areas surrounding the new parks. Birkenhead Park near Liverpool (which inspired Frederick Law Olmsted) was in large part a real estate venture, as were Regent's Park and Hyde Park in London—and so, for that matter, was Central Park, as Olmsted acknowledged. It should be recognized that over the last two centuries the large public park has helped accomplish these objectives: the park is a source of health and pleasure, it is a work of art, and it has had a powerful influence on the evolution of the city. We are now aware that its role is changing, however, and that the resurgence of the vernacular resort, thanks to the automobile, has diminished the park's overall importance in the lives of many urban Americans.

The typical post-Olmsted park, Idaho Falls, Idaho. (Photo: Barrie B. Greenbie)

Of late, the Olmsted tradition has been the subject of reappraisal: while his work as a landscape architect and urbanist is almost universally admired, his social philosophy has been criticized because of its elitism, its anti-urban tone, and its overemphasis on the moral impact of the natural environment. Most important, his definition of the park called for the fragmentation of society, for the solitary experience or at least for the family experience, and for a passive relationship between the individual and his or her environment. Roger Starr has formulated this last objection in a telling paragraph: "Most people . . . recognize that the non-human world is a battleground for survival, and that if it can be said to set forth any example of morality, it does so by providing an escape from those human vices connected with speech: hypocrisy, lying, group hatred, envy. It is a long list. What the park offers is a relief from the constant buzz of the city, and the stony messages implicit in its buildings. The park offers speechlessness." Starr comments on Olmsted's emphasis on the "tranquillity" of the park: "The word . . . suggests not merely silence but peace, the absence of contention. . . . The park designer, like the poet [Wordsworth, in his sonnet on the view of London], saw in the silent speechlessness of stone and buildings a moral orderliness that the weekday city could not provide."[3]

What we now see is the proliferation of ad hoc public spaces where the interaction and confrontation of the marketplace prevails: the flea market, the competitive sports event, the commercial street in the blue-collar part of town temporarily transformed into a fairgrounds, the parking lot transformed into spaces for games and spectacles, and the popularity of those small downtown spaces that William H. Whyte has so effectively studied, where the presence of others is the main source of pleasure and stimulation. The park as a total experience for all classes of citizens is gradually becoming merely one space out of many, now serving an invaluable function primarily for children, older people, and the dedicated student of nature, while the more mobile, more gregarious elements seek recreation in shopping malls, in the street, on the open road, and in sports arenas. Do they find it? Not as often as they should; and those ad hoc spaces are the ones to which the designer/landscape architect or developer or architect should give form and meaning.

9 Vernacular Gardens

A renegade front yard. (Author's photo)

As now practiced, gardening is essentially a solitary occupation, and we explain our liking for it in terms of an individual experience: it is politically correct to say that we enjoy gardening as a kind of therapy or as a way of expressing ourselves, and it is environmentally correct to call it a way of getting close to green nature.

But the moment we introduce the house, the home, into the picture, the garden becomes a different thing, and gardening assumes an *ethical* quality: we are working in that small plot of land because tradition tells us that we ought to provide for the family; house and garden form a definite unit: each needs the other, and gardening is a group undertaking.

We assume, quite incorrectly, that the enclosed family garden, the area set aside near the house for the cultivation of plants and vegetables and trees, is a worldwide phenomenon—even though the Garden of Eden was definitely not of that kind. In fact the enclosed family garden is unknown in subtropical regions of Africa and Asia, and it is by no means the only kind of garden we have here in America. Geographers and agriculture historians tell us that the family garden associated with the house is confined to certain cultural landscapes, and that even in our own American-European world it probably came into existence no longer than about two thousand years ago. It has an ethnic origin and an ecological origin, a fortunate combination which may account for its persistence.

In a subtropical landscape with a great diversity of trees and plants and fertile soil, farmers plant their fruits and vegetables in a field well outside the homestead, and sometimes along the banks of streams or at random on the edge of the forest. That is because those plants are native to that environment, even though they have been domesticated: they are at home, as it were, and need very little care. The land around the house is used for a variety of purposes: for holding cattle and domestic animals, for work and sociability, but *not* for a family garden.

A medieval peasant garden.

Our own domestic plants have a different relation to their environment. Our Indo-European ancestors moved out of Asia some eight thousand years ago and, after much wandering, settled around the Mediterranean, in Russia, and in northwestern Europe. It is from that last group that we derive our garden tradition.

Those barbarians—to use the term as meaning non-Greek or non-Roman—practiced a primitive kind of agriculture: raising livestock (cows, sheep, pigs, horses) that grazed in the forests and in the open grasslands; and they also planted small fields of grain. Since the soil in northwestern Europe around the Baltic, the North Sea, and facing the Atlantic was poor, the fields were only temporary. The region had been so scoured by glacial activity that it lacked in botanical diversity. The newcomers found few edible fruits or plants and had to rely on the small stock of seeds and cuttings they had brought with them from their homeland or had accumulated in the course of their wanderings: onions, turnips, cabbage, beans, lentils, and others. Few of these could easily adjust to the soils and the climate of northern Europe, unlike the domestic plants of subtropical landscapes. They therefore had to be protected and kept in a special enclosed space. The word *garden,* like the Latin *hortus,* derives from an Indo-Germanic root meaning fence or enclosure.

In prehistoric times, and probably well into the Middle Ages, that garden or farmyard played an important role in the life of the family: as garden, it was where those vegetables were grown, along with medicinal and magic herbs and fruit and nut trees. The fruit trees provided calories as well as juices. The nut trees gave oil; and somewhere in the garden there was likely to be a beehive. But the garden also had symbolic value: to plant a tree was a sign—and it still is with many farm families—of settling down, of taking possession of a piece of land. That is why the garden, in almost every European landscape, had a special legal status: it was the primary *private* land: land which the community could neither tax nor expropriate, land which was under the control and maintenance of the woman of the house. One of the oldest of Roman laws stipulates that the garden is the woman's domain.

Seen as a wider space, as a barnyard, the garden was the focal point of family life. It was where chickens and geese and dogs and cows and sheep were held. It was where the men worked, mending harness or doctoring a sick horse, or unloaded and stored the harvest. It was where the women watered and trimmed and weeded the plants and provided them with an artificial environment: protecting them against wind or cold or insects; it was where some member of the family twisted rope out of the hemp grown in the garden, or wove linen from the flax.

Elsewhere (in *The Necessity for Ruins*) I have had something to say about the ethical and moral significance of the peasant garden north of the Alps.[1] By modern standards it was unsightly, dirty, and in the background stood a house with a thatch roof, pigs and chickens running in and out of the kitchen. I doubt if the men and women who lived there had much of an eye for beauty or an ear for conversation. In the words of Thomas Hobbes, their lives were "poor, nasty, brutish, and short."

Nevertheless, it is certain that their house and garden signified something more than food and shelter. Gardens teach us more than we are aware of. The men and women constantly taking care of animals and plants learned new skills, developed new tools, a new sense of time. In the long run, the garden became a microcosm and revealed the seasons and the weather and the difference in soils. It was where the members of the family worked together and learned traditions and beliefs. What would otherwise have been an endless succession of chores and responsibilities was coordinated into a routine, a schedule, a calendar punctuated by celebrations and new beginnings. It was especially the role of the women to link the house with the garden and out of the many small lives—vegetable and animal and human—under their charge, to create an autonomous community with its own customs.

The family vernacular garden (as distinct from the farmlands) underwent many changes. There is no need to repeat the story of how, about four centuries ago, those ancient, time-honored ways of working and building and planting began, in one province after another, to shift. Sparked in large measure by the growth of cities and their needs, and by the introduction of unfamiliar useful plants from Asia and America that had to be acclimatized, a different landscape, more colorful, more spacious, and more productive, came into being. Those new edible plants for the urban market had to be produced in large, uniform quantities, and to be of uniform appearance. This was when the painstaking systematic horticulture developed by the housewife in the family garden was introduced to the plowed field. Carrots, beans, onions, new plants such as corn and squash and potatoes, were planted in rows, weeded and cultivated and fertilized, then harvested and promptly carried to market.

One aspect of that earlier agricultural revolution and of the transformed landscape was the wave of rebuilding in towns and in the countryside. Growing cities were beautified with broad streets and masonry houses with tile roofs and handsome facades. Many villages, destroyed in wars or allowed to decay, were redesigned by their landowners. In the sixteenth century England passed laws to insure that every new rural dwelling had at least four acres of land—an area too small for plowing and farming in those days, but large enough for an ample family garden. Similar laws were passed in parts of France and Germany. The artificial military settlements on the frontiers of eastern Europe consisted of rows of identical small houses, each with its own enclosed garden space. They are still much in evidence in Hungary and Poland.

One peculiarity is worth noting: in almost every case the house with its garden was at some distance from the fields and meadows where the men worked. It was as if the ancient distinction between house and garden on the one hand, and open fields and grazing land on the other, was still honored: as if the garden was still seen as having a special status.

It was a beautiful and prosperous landscape, with wide, evenly plowed fields, pastures enclosed by hedges, new houses and churches, roads lined by rows of fruit trees. In the country, merchants and government officials built impressive mansions with lawns and ornamental gardens, and the new techniques of growing fruits and flowers and vegetables transformed many villages and farmyards into abundant displays of fruit trees and rows of vines. It is not hard to find scholars and architects and painters who declare that the sixteenth and seventeenth centuries was the period when our traditional Western landscape came into full flower, as if it too were a garden, lovingly cared for by its inhabitants.

Yet there were critics even then who reminded us that in that most elegant and picturesque of landscapes many thousands of men and women wanted nothing more than to sail to the New World and find a better way of living. They came with the intention of producing an improved version of the ancestral landscape and the ancestral economy. The layout of many of the early colonial villages resembled the prototypal model: a cluster of dwellings each on a plot of several acres near to, but distinct from, the family fields. Further away lay the village grazing lands and the wilderness forest, seen as the area of potential expansion. In the center of the community were the church and the village green or common. It seemed an arrangement well suited to farming and the raising of livestock, and to community existence. There was more than enough land, and the possibility for each settler to acquire more.

The first years were inevitably hard. The supplies of food the settlers had brought with them did not suffice, and the forest environment, new to those who came from towns or open landscapes, was intimidating. They set to work building houses, clearing land for fields, and survived, confident that they would soon be raising crops and that their small stock of cows and pigs would soon increase in number.

In the Old World forests had been reduced and overexploited, but in America they were not only immense but rich in untouched resources: fruits and berries far larger than any the settlers had previously seen, grapevines and nut trees, much game, and an inexhaustible supply of timber. There were trees in New England which in the spring gave generous amounts of a sugary sap. As they explored their new environment, the settlers discovered many wide clearings in the forest, some as large as five hundred acres, where all trees and undergrowth had been removed. These clearings had been made by the Indians for their crops of corn and beans and squash and pumpkin. The more adventurous settlers took possession of the open land, deserting the still unfinished villages.

From friendly Indians the settlers had learned that those native vegetables were not only delicious and nourishing, but extraordinarily easy to plant and raise: the soil had only to be scratched with a hoe. Moreover, since they were native to the region, they needed no special care. So here was a reliable source of food which grew almost everywhere, calling for no enclosed garden. Preparing land for wheat was an arduous undertaking, and the crop proved susceptible to disease. Corn, on the contrary, produced an excellent flour. In short, the New World offered a diet which made the traditional garden and traditional garden work unnecessary.

The settlers were further elated to discover another plant grown by the Indians that had a commercial value: tobacco. In the southern colonies, the raising and selling overseas of tobacco occupied the thoughts and energies of all, often to the exclusion of any plans for gardening. Indeed, the space set aside for gardens in the homestead was often taken over by tobacco; tobacco was even grown on the roads. The trouble with growing tobacco (and to a lesser extent with growing corn) was that it soon exhausted the soil, and fresh fields had to be found elsewhere; this meant that the homestead had to be moved. But land was cheap, and after the first generation moving became a way of life for many young and ambitious families, already restless in the established, somewhat regimented villages nearer the coast.

Under these circumstances, who (aside from the women) had time to give thought to a garden? From the beginning there was a great shortage of laborers: wages were high and the back country had a strong appeal. The men of the family, as was the ancient custom, labored in the fields, producing or extracting crops for export—corn or tobacco—or cutting timber and collecting fur pelts. They had little interest in improving the home grounds. "They dug up and fenced a small piece of ground near their houses," says an agriculture historian, "and left its care to the women of the household, already overburdened with household industries"—spinning, weaving, cooking, baking.[2] Except in the older villages of New England, the average colonial homestead was isolated. Being a combination housewife and gardener was not only hard work, it was lonely. In the Old World gardening had meant having neighbors and learning from them, and in the Old World landscape there were always

Many medium-sized farms devote to ornamental planting a square plot of land exactly in front of the dwelling. (From *Rural Affairs,* 1873)

examples of quality and expertise to imitate: the village priest, who knew more about pruning than anyone else, and the vegetable garden of the lord of the manor, maintained by four full-time gardeners. Though no peasant wife could dream of copying them, they set a standard and introduced new methods. But pioneer landscape offered no such stimulation. Gardening was monotonous, and the farmer himself did little to encourage a kind of work which seemed of little consequence to the farm as a whole.

Something like a faint remnant of the ancient belief that the garden was the domain of women lingered in colonial America, but it was condescending in tone. Samuel Deane, author of one of the first books of American agriculture, published in 1797, wrote that "farmers as well as others, should have kitchen gardens, and they should not grudge the labor of tending them, which may be done at odd intervals of time which may otherwise chance to be consumed in loitering."[3] And even that offhand help was rarely forthcoming. Long after large-scale commercial farming had become general, the woman of the home persisted in gardening, though alone and without encouragement. "The country is full of rural improvements," said the periodical *Rural Affairs* more than a century ago, "but the improvement of farmers' gardens is not one of them. . . . We have just driven past the garden of one of the best farmers in the country. . . . It was hard to say whether weeds or vegetables had the ascendancy. . . . There was an evident struggle between a refined taste on the part of the woman and utter neglect on the part of the man."[4]

It is ironic that while throughout the eighteenth century in America many prosperous urban families possessed attractive ornamental gardens and employed full-time gardeners, equally prosperous farmers found it easier to buy vegetables on the market or to rely on the corn and beans from their fields. Only one kind of garden in those days could be said to resemble the classical vernacular garden in the sense that it produced food to supplement the field crops and was a family undertaking: the garden of the slaves on many large southern plantations. Families were sometimes allotted land back of their house, and there they raised vegetables and herbs and fruit from seeds, many originally brought from Africa. They thereby relieved the monotony of the rations provided by their masters and kept alive a fragment of their heritage. Ethnobotanists are now discovering the variety and the value of the plants grown in the slave gardens and still propagated in many black gardens—an instance of how the house-and-garden complex can serve to protect and foster a culture which is impoverished and besieged.

But like the vernacular garden of peasant Europe, the vernacular garden of early pioneer America was eventually rendered obsolete by commercial farming. The homestead became an element, and not always an important one, in a nationwide system of agricultural production and distribution. It is of course pleasant to see, in the working-class sections of our cities, small gardens strikingly different in design and materials from the average front yard of lawn and shrubbery: they are the creation of recently arrived minorities; but sooner or later, as their owners are assimilated, those gardens will vanish.

What happened to the space previously devoted to the kitchen garden in workaday America? Did a new kind of vernacular garden make its appearance? This is a question of some interest to those studying the evolution of the American landscape, but so far landscape architecture historians and Americanists have given it little or no attention. One reason is that many of those involved in landscape architecture—either as professionals or as teachers and writers—are convinced that for reasons of prestige the design of gardens should be treated exclusively as an art. The social and economic roles of designed outdoor spaces, parks, freeways, gardens, nature preserves are not always taken into account, and the only gardens worth examining are those whose designers belong somewhere in the apostolic succession starting with Capability Brown.

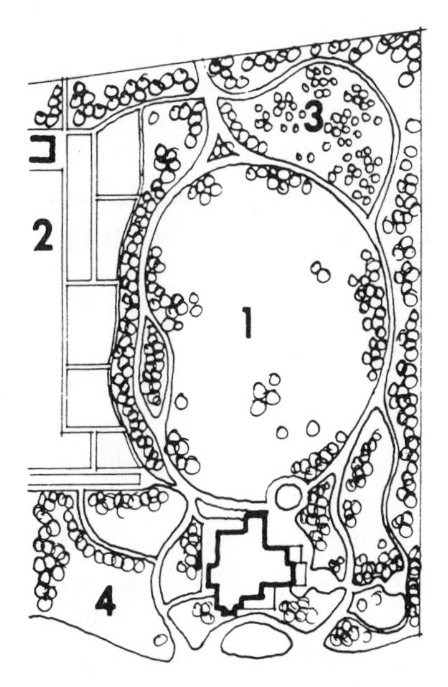

(top) The popularization of the A. J. Downing style. (From *Rural Affairs,* 1873.) (bottom) Downing's plan for the suburban villa garden, showing lawn (1), kitchen garden (2), picturesque garden (3), and flower garden (4).

Ask any landscape architecture historian what style evolved out of the pioneer or colonial garden in America, and he or she will answer without hesitation, "The style of Andrew Jackson Downing, who was a disciple of Humphrey Repton."

Downing was a fashionable (and talented) designer whose clients, to his great satisfaction, were chiefly rich and sophisticated families eager to impress their friends with their money and good taste. But the great majority of Americans who were building houses and laying out farms in the mid-nineteenth century had only heard his name. "Notwithstanding the high character and adaptability of Mr. Downing's work to the 'Upper Ten Thousand,'" Solon Robinson wrote in 1845, "the wants of the lower ten hundred thousand are not satisfied."[5]

What was their solution? To a pioneer farming family settling in Indiana or Iowa, the problem was not acute: clear away the dense growth of trees, lay out four rectangular fields, build a house of logs, a barn, fences, and plant a profitable crop. All that this entailed was a clear space around the house and an organization of work spaces. But it was a *new* organization: barns, sheds, storage facilities, livestock being separate from the living quarters; and it meant that the dwelling was seen as temporary, that in a few years it would be replaced by a comfortable and sightly house with a porch and a view. Judging from the accounts of travelers, the new settlers marked time until they were prepared to beautify the home.

The early nineteenth century was a time of experimentation in the landscape, and of makeshift solutions. The best evidence we have of how the immediate environment of the house was treated comes from such ephemeral literature as the catalogs of nurseries and farm equipment firms. There is also much to be learned from state and federal publications on farming. A number of family journals, well illustrated and full of advice, give us glimpses of landscaping problems, and travel accounts, though dwelling largely on real estate deals and land prices, are often revealing. Immigrant handbooks describe the pioneer countryside from the point of view of the newcomer short of cash, short of experience, but eager to make a home. None of these sources has much to say about the ornamental garden: the emphasis is on fruit trees, grain crops, livestock. It is only in the books on how to build inexpensive houses, using the new balloon frame technique, that we run across any mention of the aesthetic needs of the families.

A front lawn that avoids the suggestion of work or productivity. (Photo: Barrie B. Greenbie)

When it is a matter of defining a neovernacular style in America, the abundance of popular material in the twentieth century is overwhelming—though not in the establishment landscape architecture publications. Yet what makes the investigation rewarding is the amount of *visual* material now available in every suburb, in every residential street, and even in farmsteads. What the contemporary professionally designed garden seems to have in common with the vernacular home-made garden is the rejection of any suggestion (at least in the front yard exposed to public view) of work or productivity. In the Downing-designed landscape there is always an attempt to conceal or disguise such utilitarian structures as the stable, tool shed, the house of the gardener or coachman, and it is scarcely necessary to say that any trace of the plow, of a vegetable garden or orchard, is hidden. This would be a logical omission in a private landscape devoted to trees and lawn with romantic glimpses. What is more remarkable, however, is the equivalent rejection of work or the suggestion of work in the gardens or front yards of much more modest wage-earner houses. If the prosperous suburban household seeks to eliminate all evidence of work and productivity from its surroundings, it is because its ideal is *environmental:* the garden (or the grounds) as an oasis of untouched nature. On the other hand, the less affluent homeowner eliminates all evidence of work in order to promote an atmosphere of family leisure and *recreation.*

This tendency toward a garden designed for play and relaxation has taken over in most modern urban and suburban households, and even in trailer courts. One hears of developments—middle-class as well as those which are expensive and exclusive—where covenants prohibit the parking of work vehicles in the driveway, or working on the car in public, or neglecting to mow the front lawn; all commercial signs, all references to a workplace or a service offered are forbidden. One of the more oblique ways of repudiating the notion that the garden or front yard could be a place of work is the casual display (as ornaments) of obsolete farm equipment: a plow, a wagon, even an old-fashioned hand-operated lawn mower. All make the point that work is out of date. By the same token, there is no more unmistakable sign of poverty and indifference to standards of respectability than the presence in the front yard of cars being worked on by the younger generation.

The evolution of the yard or garden from a space essential to the welfare and cohesion of the family to the sterile display of close-cropped grass and foundation greenery is worth study. The notion that the garden is now perceived either as a microenvironment designed by a full-time ecologist and harboring endangered species, or else as a place of leisure, is not easy to accept. But those are the two versions—the one establishment, the other vernacular—that are beginning to prevail. For there is no prototypal garden *design:* there is only the prototypal *garden* identified with house and family; and these contemporary versions will eventually take their place in landscape history.

(Author's photos)

Yogi's
Hand
Car Wash

10 Working at Home

Garage, New Haven, Connecticut. (Photo: Tom Strong)

When should we keep the place of work separate from the place where we live? It depends very much on the kind of work. In the city, factories and heavy traffic make certain areas all but uninhabitable, and we protect residential neighborhoods from contact with industry by means of zoning. But in a small town or a village the problem is more complicated: we want to preserve the green, quiet atmosphere of our residential streets, yet we are reluctant to exclude families who depend on a home enterprise. I have neighbors who work in town all day and whose houses and front lawns are models of small-town domesticity; but I also have neighbors who operate a laundromat and others who live above their machine shop. Their front yards are disheveled parking lots. Still, I enjoy doing business with them. They are near at hand and they are friendly.

What is at stake in this and similar instances is a matter not so much of aesthetics or property values as of how we define the home and its role in the community. That is a definition hard to come by. I thought I had found the answer in a publication called *Home: A Place in the World.* It consisted of the proceedings of a conference held in 1990, attended by a number of social scientists, historians, architecture critics, and other authorities. In the words of the editor, the conference was designed to "explore the ideology of home, its meaning as a central idea, as well as the crises engendered by its loss in homelessness and exile, and by the experience of loss suffered in alienation."[1] An impressive agenda!

The book opened my eyes to the complexity of a subject which I had thought I understood. What the speakers discussed, often with eloquence and learning, was the *idea* of home, home as an individual, sometimes solitary experience. The notion of being at home, for instance, was defined as "a mental and spiritual condition," and Georg Simmel was quoted to the effect that "home is an aspect of life and at the same time a special way of forming, reflecting and interrelating with the totality of life." I learned that home could be likened to a set of Emersonian conceptual concentric circles.[2]

I also noted, to my surprise, that house or shelter actually had very little to do with home. There were disparaging references to the current use (or misuse) of "home" as the equivalent of "residence"—"the linguistic waste product of the real estate industry." Certainly the joys of returning to the homestead have often been exaggerated, but I was struck by the fascination which the concept of homelessness seemed to hold; no less than four speakers expatiated on what was termed "a somber and significant terrain," and one speaker declared that the real alternative to homelessness was "not shelter but solidarity."[3]

This outspoken hostility to the house as one aspect of home was puzzling. Some of it was clearly inspired by an urge to astonish, to shock; but I began to understand the attitude after reading in one of the papers a reference to home as a withdrawal into the safekeeping of our dwelling. "The cloister and the cell as home, places of meditation and work are reflected in secular modernity by the idea of the writer's home . . . to which one retires from the outside world or family, bed, and board of the rest of his house."

So the cat was at last out of the bag. Despite all the discourse about alienation and exile and the grandeur of homelessness (especially for the writer and thinker), home proved to be little more than an academic version of the middle-class American house dedicated to privacy, leisure, and remoteness from the workaday world.

"Western culture," Yi Fu Tuan has written, "encourages an intense awareness of self and, compared with other cultures, an exaggerated belief in the power and value of the individual. . . . This isolated, critical and self-conscious individual is a cultural artifact. We may well wonder at its history. Children, we know, do not feel or think thus, nor do nonliterate and tradition-bound peoples, nor did Europeans in earlier times."[4]

Yi Fu Tuan noted that in the evolution of the European house "more and more rooms were added that enabled the householder and his family to withdraw from specialized activities and to be alone if they should so wish. The house itself stood apart from its neighbors." He mentioned the various ways in which the middle-class or academic householder withdrew from the public sphere: by a complete rejection of gainful employment in the home, by a sentimental cult of closeness to nature, and finally by a clear-cut, unmistakable separation of the residence (in the suburbs or in exurbia or in the condominiumized wilderness) from the office or factory or classroom.[5] I find that the notices of houses for rent in the columns of the classifieds in the *New York Review of Books* or in the *Nation* give a wonderfully concise description of the ideal home of the professional or academic citizen: "Charming, secluded, environmentally friendly house: three bedrooms, three-car garage, swimming pool, solar energy, extensive library, breathtaking views of unspoiled rural landscape. Ideal for sabbatical hideaway or nature contacts. No smokers need apply; no pets, no children."

There is much comfort in the thought that this decadence is confined to a very small class, and that now, as in the past, the vast majority of Americans are committed to a very different definition of the home. As one of the speakers at the conference observed, "Most historians have tended to generalize for the whole society on the basis of the middle-class experience. The process by which working-class families eventually adopted the new domestic lifestyle has not been documented. . . . For working-class families the home was not merely a private refuge; it was a *resource* that could be used for generating extra income."[6]

The academic and professional middle class want their house to be as inconspicuous as possible, to avoid being ostentatious, and to blend with the natural environment. But for most of the rest of us, the house is there to be seen. It shows that we are permanent members of the community—village, neighborhood, parish, school district, subdivision. In the words of a philosopher, "Property makes a man visible and accessible. I cannot see a man's mind or his character. But when I see what he has chosen and what he does with it, I know what he likes, and quite a good deal about his principles."

What the average contemporary American dwelling tells us about the family is whether it is rich or poor and how much it values public opinion. It tells us nothing about how it makes its money, and reticence on that score is one of the benefits of our emphasis on privacy. But until three or four centuries ago in Europe, the size and exterior features of the house told us the social status of the family and how it contributed to the community; that was because in those times home and place of work were one and the same. This was even true of the house or castle of the nobleman: by law he was allowed to adorn it with castellations and a moat and a dungeon to indicate that he had juridical powers and was committed to defending the community. The number of bays in the house of the yeoman indicated the number of acres he farmed and what he paid in taxes; in the case of certain ancient homesteads, a seven-foot fence around them showed that the owner had the right to maintain the "King's Peace" among his servants and in his family, without police interference. The lowliest of houses was the one-bay "cottage" with less than enough land to farm. The cotter (or occupant) supported himself and his family by working for others and by what we now call cottage industries: the producing of everyday items—tools, pots, harness, even food—which the other villagers could buy.

Thus almost every house in a medieval village participated in the life of the community—as a place of work or of certain services. No less universal was the emphasis on visibility and accessibility. The cottage was open to the buying public and to the authorities; even the nobleman's house had its hall for public assembly and its court for trials. It could be said that community flourished at the *expense* of privacy, not to preserve it. In towns where space was limited, the absence of privacy was notorious. A family and its hired help often lived and worked in one room, and much of their activity spilled over into the street, where they displayed their wares. If the location of a house inconvenienced the flow of traffic, or even if it was the scene of too much rowdiness and noise, it could be moved or destroyed. Noble families were not exempt: they were obliged to build houses which were appropriately large and ornate, with an imposing facade on the street.

The community organized around work and public service seems to have functioned most smoothly in rural villages where the rich and powerful stayed on their estates, and the farmers, already accustomed to producing for their own daily needs, set up their own home industries and made money selling to the villagers—indeed, many of them came to consider their farm work merely a secondary source of income. In *The Colonial Craftsman,* Carl Bridenbaugh tells of how many colonial villages, especially those in New England, rapidly evolved their own group of basic home manufactures and crafts, located within or next to the dwelling: part-time farmers who produced wagons, tools, utensils, tanned leather, made hats and shoes and furniture, and even produced food—"to the great convenience," in the words of an eighteenth-century commentator, "and advantage of the neighborhood."[8] What we forget, in our admiration of the colonial village, is that it long retained those medieval controls on the size and location of the house, the limitations on privacy, and collective work obligations.

This arrangement came to an end with the industrialization of many crafts in the latter half of the eighteenth century; first in the towns, then in the countryside. Thomas Hubka's book *Big House, Little House, Back House, Barn* is a remarkable study of how many New England farmers sought to keep alive the traditional relationship between home industries and the community, only to succumb to a new market-oriented, one-crop agriculture. But in terms of the house the divorce from community control and from the workplace came much earlier. Yi Fu Tuan gives instances of it in the fourteenth century. Philippe Ariès dates from the sixteenth century the house of the merchant and prosperous farmer designed as a private, autonomous domain dedicated to the joys of family life.[8] It was only in the nineteenth century, however, that the average American family discovered privacy in the home. The monotony and shabbiness of many company towns and tenement buildings and early subdivisions should not deceive us: each house is a private refuge; all suggestions of community and of the job are remote and invisible.

This is by no means the end of the story. Architecture historians, concentrating almost exclusively on the evolution of the middle-class house, avoid discussing changes in the wage-earner house over the last fifty years, and social historians discuss the place of work largely in terms of the factory or mine or corporate farm. The ancient tradition of working at home as a secondary source of income is either ignored or dismissed as a kind of tinkering—made fashionable as a topic by Lévi-Strauss's writing about bricolage. Someday a student will discover the American tradition of home industry, first as it expressed itself in the use of wood—a craft which nineteenth-century European travelers much admired—and then in the mid-nineteenth century in our mechanical skills. It was on the farm that these were first manifest, and to this day the farmer is still an inventor of labor-saving devices and ways of using power.

The urban worker, lacking space at home and the expensive tools necessary for mechanical work, only really found an outlet with the popularization in the 1930s of the low-cost family automobile, closely followed by the popularization of the truck (and other commercial models) for family-oriented work. Possession of this expensive and useful object involved not only repairs and maintenance, but improvements and ex-

A barber working at home, Cordova, New Mexico, 1939. (Courtesy of the Museum of New Mexico, Santa Fe)

perimentation, and a new money-making career evolved—always centered on the house—of hauling and distributing and collecting, and of transporting passengers, usually on a small, local scale. Though the house itself was left inviolate by this new home industry, the front lawn, the backyard, and the margin of the street were all taken over, to the dismay of neighbors. Further developments ensued: after World War II almost every low-cost house had an attached garage—spacious, equipped with light and power, easily accessible and very visible. It provided space for work and for keeping tools, and its open door and driveway encouraged neighbors to come by and offer advice. Furthermore, it liberated the house itself from the dirt and confusion of the workplace and the occasional appearance on the kitchen table of oil filters and orange rags. The garage, in short, restored something like the old order of things: work in one part of the house, privacy in another.

One of the less celebrated accomplishments of technology was the production, beginning (I believe) in the 1950s, of power tools for the home. Power tools in industry and in construction were already common, but their availability in stores or for rent gave a remarkable boost to every garage industry and private craftsman. It is true that when we take the trouble to explore a blue-collar neighborhood we are struck first of all by the immense number of garage industries focused on the automobile. They transcend all zoning regulations, all preservation programs, all ethnic barriers—except in the most regimented of planned neighborhoods—and bring with them a scattering of used car lots and auto junkyards and gas stations, and roads and streets clogged with traffic. But when we note the other, less spectacular home industries, we discover that they are in fact more numerous.

If these have any common denominator, it is that they do chores and provide services that the modern family has neither the time or talent to cope with. Even the most modest household, even the smallest trailer, contains a clutter of gadgets, most of them electronic and all of them prone to malfunctioning, from the electric carving knife to the electric trash compactor and the electric blanket. Invariably, they get out of whack long after the guarantee has expired. What to do?

There is a man on Maple Street who will take care of your problem when he gets home from work. You will find him in his garage. You will also find, on another emergency occasion, a man who can mend furniture or one who can put your power mower in shape; and elsewhere (not in the garage, but in the house) a woman who bakes and decorates birthday cakes, another who sells medicinal herbs, yet another who is a part-time babysitter or instructor in classical guitar; and a man who with his son can repair computers and work on your car radio. All of these helpers request cash payment to avoid income tax complications.

How do you find them? They never advertise, they are not in the yellow pages, and when you *do* locate them they are likely to be away. It is essential that you be familiar with the neighborhood; it is essential that you know work hours, can recognize the craftsman's car outside his or her favorite leisure-time resort—bingo parlor, laundromat, church. In a word, to take advantage of this array of industries and services you have to be a member of the community of long standing.

There are two obvious reasons why these home enterprises flourish and multiply: they are convenient for their customers and they

Repairing ten-speed bikes at home. (Photo: Tom Strong)

are profitable for their owners. As we all know, our towns and cities have expanded enormously, thanks largely to the great increase in car ownership, and as a result it is a great undertaking to go down into the central city to established service and repair facilities. The modern mall, according to conventional wisdom, is the successor to Main Street, but in fact the mall has no room in its lavishly landscaped precincts for one-man enterprises. Who, in fact, has ever seen a shoemaker or an upholsterer or a place where a wheelbarrow can be fixed, in a mall? That is one good reason for garage industries to be welcome and even essential. They are small, they are near, they and their operating procedures are visible and accessible. Their background of domesticity—children and dogs and a vegetable garden, and the smell of supper being prepared—makes the encounter a face-to-face social occasion. How can you complain when the job is less than professional and takes three days? We are all neighbors and are likely to meet soon again, at church or at the supermarket.

For the craftsman himself, the rewards are no less substantial. He is able to use the mechanical or industrial skills acquired in his full-time job to make extra money at home. He makes friends and plays a role in the local business world. If he is unusually skillful or inventive, he will be discovered by a wider clientele.

I have used the word *community* very often and, I'm afraid, very loosely. I have meant that I was interested in establishing, very roughly, the *boundaries* of a kind of working-class neighborhood where everyone is mobile, has limited leisure time and a limited income; a community where everyday domestic needs can be satisfied by the people who live nearby; by the contribution each household can make to the smooth flow of existence. A community of this sort does not derive from any utopian dream or any compact. In many instances it comes into being imperceptibly and naturally, and it seems to work surprisingly well. I attribute that, at least in part, to the way in which people in the community define and use their house or home.

Many years ago I suggested that the low-income house, whether owned or rented, whether a trailer or a bungalow, could be likened to a transformer in its effect on those who lived in it. "The property of transformers," I wrote, "is that they neither increase nor decrease the energy in question, but merely change its form. . . . [The house] filters the crudities of nature, the lawlessness of society, and produces an atmosphere of temporary well-being, where vigor can be renewed for contact with the outside."[9]

That was a definition emphasizing the privacy of the house, the interior as a refuge, and I still believe that this can be an important aspect. But the family itself, to say nothing of the public, judges the house as it relates to its surroundings, natural as well as social. We see the house as a sign not only of membership in the community, but of its interaction with the community. So I am now inclined to believe that a better metaphor for the average house is as the *extension of the hand.* It is the hand we raise to indicate our presence; it is the hand that protects and holds what is its own; the house or hand creates its own small world; it is the visible expression of our identity and our intentions. It is the hand which reaches out to establish and confirm relationships. Without it, we are never complete social beings.

TOWNS, CARS, AND ROADS

11 A Sense of Place, a Sense of Time

(Photo by Bill Owens from *Suburbia*. San Francisco, 1972)

Most foreign visitors to the United States end up liking us. It is our landscape that bewilders them and that they find hard to understand. They are repelled by its monotony: the long straight roads and highways, the immense rectangular fields and the lonely white farmhouses, all much alike. They remind us that in Europe every city has individuality, whereas in our country it is often hard to distinguish one city from another. With the possible exception of Boston and New Orleans and San Francisco, they not only are lacking in architectural variety, they are lacking in landmarks and in unique neighborhoods. We are often asked how we who live in the midst of such monotony can have any sense of place.

I find this hard to answer. Most of us, I suspect, without giving much thought to the matter, would say that a sense of place, a sense of being at home in a town or city, grows as we become accustomed to it and learn to know its peculiarities. It is my own belief that a sense of place is something that we ourselves create in the course of time. It is the result of habit or custom. But others disagree. They believe that a sense of place comes from our response to features which are *already* there—either a beautiful natural setting or well-designed architecture. They believe that a sense of place comes from being in an unusual composition of spaces and forms—natural or man-made.

In any case, plenty of thoughtful Americans see eye to eye with those foreign critics and wish that we could somehow give our downtown areas a sense of place. Much has been accomplished, in fact, in America in the way of injecting life and design into the decaying central city: the streets have sometimes been turned into pedestrian walks with brick pavements and fountains, adorned with planters and brilliant flowerbeds. Small parks with rows of trees and a piece of abstract sculpture have been inserted between the glass high-rise buildings, and many efforts have been made to conceal the original grid layout of the downtown area. There are concerts of Baroque music in the new minipark and ethnic pageants, each of them featuring the costumes and dances and food specialties of a group.

On such occasions the whole area is brought to life. A kind of invisible confetti fills the air, and we feel that the central city has at last become an exciting and stylish part of town, the old monotony banished forever. The sense of place is reinforced by what might be called a sense of recurring events.

I have recently had a chance to see what has been done in the way of revitalization in such cities as Dallas and Houston and Denver and Oklahoma City and Memphis and even Little Rock. I had the feeling that this expensive facelifting affected the rest of the city very little. Architecture buffs enjoy the results, and so do tourists, but if you are a resident of the city or merely on your way to work, you see the display in a different light.

Say you are passing through the renovated downtown late at night: you then find that the dominant feature of the scene is not the cluster of magnificent forms and spaces; it is the long and empty view of evenly spaced, periodically changing red and green traffic lights along Main Street. The tall glass buildings, so imposing by day, are half-hidden in darkness and stand to one side to allow the street to thrust ahead, unimpeded. It cuts through the less opulent parts of town, the block after block of silent, nondescript houses like the houses in every other American city. It goes through the tree-grown suburbs and parallels the complex of warehouses and parking lots and industrial plants until at last it turns into an interstate highway, heading into the dark and featureless countryside.

The highway never seems to end. There is an occasional brightly lit truck stop and the lights of a bypassed town. Rows of trucks are parked for the night at rest areas, and with the hours of solitary travel there comes a mood of introspection. A favorite episode in novels and movies and television shows laid in the American heartland is that lonesome ride through the night landscape: an occasion for remembering other times. You think back over your past, think about your work, think about your destination and about those you have left. The dashboard display shows how fast you are driving, tells you the hour and how many more miles you still have to go. The sameness of the American landscape overwhelms and liberates you from any sense of place. Familiarity

The centrifugal American city. (From "How Fort Worth Became the Texas-most City," Amon Carter Museum Publication, Forth Worth, 1973)

makes you feel everywhere at home. A sense of time passing makes you gradually increase your speed.

This all-pervading sameness is by and large the product of the grid—not simply the grid of streets in every town and city west of the Mississippi, but that enormous grid which covers two-thirds of the nation, stretching from the Mississippi and Ohio to the Pacific, from the Rio Grande to the Canadian border, beyond which it extends in a slightly modified form well into the northern subarctic forest. It is this grid, not the eagle or the stars and stripes, which is our true national emblem. I think it must be imprinted at the moment of conception on every American child, to remain throughout his or her life a way of calculating not only space but movement.

Let me say something about the impact of the grid on a part of the country that I am familiar with: the High Plains. It stretches along the eastern slope of the Rockies from Canada to Mexico, and as far as Iowa and Missouri and Arkansas to the east. It is magnificent, undulating country, and before white settlers arrived it was where hundreds of thousands of buffalo grazed on the expanse of short grass. The wind blows incessantly. The High Plains now contains many wheat farms and ranches, and when you fly over it you are struck by the miles and miles of perfectly round green fields, the product of pivot irrigation. The population is not large, and except for Denver and Omaha it has no cities with more than two hundred thousand people. None of them—again, with the exception of Denver—is brilliant or exciting, but they seem prosperous. What I particularly like are the pleasant residential districts: the streets of small, comfortable houses, each with a backyard and a well-kept lawn, each with a pickup on the driveway, streets that reach for miles and miles to where the open, treeless rangeland begins.

Since the High Plains is lacking in sensational topographical features and is wide open, it is an excellent place to observe how overpowering the national grid system can be. It is easy to suppose that when the first settlers confronted this monotonous sea of waving grass, they must have longed to divide it into squares and rectangles in order to give it something like a human scale. The round fields of their descendants suggest the same liking for simple geometrical forms, which can in fact be extremely handsome.

The grid system was already a familiar American landscape feature, almost a hundred years old when the High Plains was being settled in the decades after the Civil War. But whereas in the older states of Ohio and Indiana and Illinois (where the heritage of the colonial farm village lingered) the straight lines of the grid were valued as an efficient and democratic way of organizing individual landholdings, west of the Missouri the grid played a much more decisive role: it was the *only* practical and speedy method of organizing space. Its long-range effect was to eliminate, once and for all, the impact of tradition and traditional spaces, and of topographical factors, in the forming of the new High Plains landscape. A composition of identical rectangular spaces extending out of sight in every direction, ignoring all inherent differences, produced a landscape of empty, interchangeable divisions like the squares in a checkerboard. In the course of time these were put to several different uses and thus acquired individuality, but in this very level, very uniform terrain, there was always the temptation to consider all uses as temporary. Space, rather than land, was what the settlers bought, and it was so easy to buy, so easy to sell, that commitment to a specific plan for the future must have been difficult for many. Freedom from tradition and freedom from topographical constraints was something they had never known before.

That is probably why the early history of the region reads less like an account of pioneer farming than of endless land speculation. A man could buy a sizeable piece of land, survey it and lay out a grid of broad streets, divide the blocks into uniform building lots, and when he had advertised that land in his town was to be sold at auction on a certain day, he had only to wait for buyers. They soon came. Someone opened a store, families built small houses on the lots, and a church was erected. But the attraction of other vacant spaces was strong, and the inhabitants of not a few of those new towns or villages moved their houses a few miles distant to be near a new railroad line—or the promise of a railroad line. When this happened, all that was left of the original town site was the empty store, the empty church, and the grid of broad, empty streets quickly merging back into the wind-blown landscape of grass.

This freedom to move from place to place and to use space as you saw fit, determined—and still determines—much of the planning and architectural design throughout the West. In the early days when everyone had or could have land, there seemed to be no reason for setting any land aside for such a specialized purpose as public enjoyment, and except for a vacant block called a park, no space was given a permanent public identity. None was dedicated (in the legal sense of the word) to community needs. Even the dwelling was thought of in temporary terms. When settlers eventually had enough money to build substantial houses, the plans and designs were simple and easy to execute. At the back of the builder's mind was the notion that in time he and his family would sell the homestead and move away. Farm journals and agricultural bulletins and emigrant handbooks cautioned settlers against making their houses too personal, too individual, lest they be hard to sell. Modest though most of the dwellings were, they were sturdy and efficient and without pretense, and a cluster of those plain angular forms, successfully defying the wind, the bitter cold, and the overall horizontality of the landscape, is a spectacle you are likely to remember: it was and is a unique regional style in the sense that it rejected any of the characteristics of the environment, of the natural region.

We are fortunate in having an abundance of pictures and plans as well as first-hand descriptions of many of those mid-nineteenth-century western towns. The best way to study the material is by comparing it with the plans and pictures of earlier eastern towns. Even before the Revolution there were a number of grid towns in the colonies, but their grid layout was in the nature of a symmetrical design, a way of giving the plan a symbolic balance and unity. As the settlers moved farther west, the grid layout gradually ignored that concern for composition, and we notice the disappearance of dedicated public spaces—the central green or square, the marketplace, the drill field. By the end of the nineteenth century the average western town was likely to contain a small park at most, and its chief public space was its busiest commercial street—Main Street, or the street leading to the depot. As the economy and the population of the town underwent changes, some other street would usually become more important, and the solidly built-up row of stores and banks on Main Street was left to decay.

The tradition of a central green or square is a very old one, and contemporary planners and architects and preservationists try to keep it alive. But it really functions as a public gathering place and symbol of unity only when the town knows how to use it. The population of many American towns comes from all over the United States and Europe. When each ethnic element has its own church, its own kind of employment, its own idea of public life, the central square means very little. A flexible and frequently shifting arrangement of streets and spaces, adjusted to new real estate values and an increased traffic flow, becomes general: residential quarters, instead of grouping around the business section, tend to move out to where the immediate future is more predictable.

So the dwellings are thinly scattered. There is inevitably something like a concentration of the older homes near Main Street and the depot. Houses, most of them small but some of them sizeable, stand at lonely crossroads or by themselves on an otherwise vacant block. You wonder what made the family choose to build so far from neighbors and the center of town: Were they perhaps squatters? Were they anticipating the expansion in real estate activity? Were they small-farmers? Any of these explanations may have been possible. Most of us, I think, have been taught that the ideal settlement pattern is one which is compact and clearly defined. A tight composition of streets and houses and spaces, with something like a landmark in the center, is generally considered normal and desirable: it is more picturesque, it is easier to control and, in earlier times, easier to defend. It encourages social interaction and it produces a colorful street life; it is convenient for pedestrians. It has a sense of place. All this is true, but it is hard to ignore the widespread evidence that many people, perhaps the majority, prefer to live at some distance from neighbors and community institutions. The establishment does what it can to keep us together. In colonial New England the church authorities forbade anyone to live more than a mile from the meeting house. Yet people continued to move out, and Captain John Smith complained that the very first colonists in Virginia wanted to abandon Jamestown and settle far from neighbors.

The truth is, Americans are of two minds as to how we ought to live. Publicly we say harsh things about urban sprawl and suburbia, and we encourage activity in the heart of town. In theory, but only in theory, we want to duplicate the traditional compact European community where everyone takes part in a rich and diversified public life. But at the same time most of us are secretly pining for a secluded hideaway, a piece of land, or a small house in the country where we can lead an intensely private nonurban existence, staying close to home. I am not entirely sure that this is a real contradiction. While we agree that scatteration and the dying central city are both of them unsightly and illogical, we also, I think, feel a deep and persistent need for privacy and independence in our domestic life. That is why the freestanding dwelling on its own well-defined plot of land, whether in a prosperous residential neighborhood or in impoverished urban fringes, is so persistent a feature of our landscape. That is why our downtown areas, however vital they may be economically, are so lacking in what is called a sense of place.

"Sense of place" is a much used expression, chiefly by architects but taken over by urban planners and interior decorators and the promoters of condominiums, so that now it means very little. It is an awkward and ambiguous modern translation of the Latin term *genius loci*. In classical times it meant not so much the place itself as the guardian divinity of that place. It was believed that a locality—a space or a structure or a whole community—derived much of its unique quality from the presence or guardianship of a supernatural spirit. The visitor and the inhabitants were always aware of that benign presence and paid reverence to it on many occasions. The phrase thus implied celebration or ritual, and the location itself acquired a special status. Our modern culture rejected the notion of a divine or supernatural presence, and in the eighteenth century the Latin phrase was usually translated as "the genius of a place," meaning its influence. Travelers would say that they stayed in Rome for a month or so in order to savor the genius of the city. We now use the current version to describe the *atmosphere* to a place, the quality of its *environment.* Nevertheless, we recognize that certain localities have an attraction which gives us a certain indefinable sense of well-being and which we want to return to, time and again. So that original notion of ritual, of repeated celebration or reverence, is still inherent in the phrase. It is not a temporary response, for it persists and brings us back, reminding us of previous visits.

So one way of defining such localities would be to say that they are cherished because they are embedded in the everyday world around us and easily accessible, but at the same time are distinct from that world. A visit to one of them is a small but significant event. We are refreshed and elated each time we are there. I cannot really define such localities any more precisely. The experience varies in intensity; it can be private and solitary, or convivial and social. The place can be a natural setting or a crowded street or even a public occasion. What moves us is our change of mood, the brief but vivid event. And what automatically ensues, it seems to me, is a sense of fellowship with those who share the experience, and the instinctive desire to return, to establish a custom of repeated ritual.

I realize that this definition automatically excludes many localities which a careless use of the term endows with a sense of place. I think it is essential that we *do* exclude many current usages. But to return to the American scene, particularly to the average American western town or city, I would say that for historical reasons few of them have structures or spaces which produce any vivid sense of *political* place. What until very recently we have had are spaces and events related to the *family* and the small neighborhood group. By that I mean not merely the home—which in the past was the basic example of the sense of place—but also those places and structures connected with ritual and with a restricted fellowship or membership—places which we could say were extensions of the dwelling or the neighborhood: the school, the church, the lodge, the cemetery, the playing field. Ask the average American of the older generation what he or she most clearly remembers and cherishes about the home town and its events and the answer will rarely be the public square, the monuments, the patriotic celebrations. What come to mind are such nonpolitical, nonarchitectural places and events as commence-

"When there are seventy people for dinner you have to be organized. We have carving and serving committees, games, and a family quilt everyone sews on. A silent auction pays for the three turkeys. More families should be proud of their heritage." (Photo by Bill Owens from *Suburbia*)

ment, a revival service in a tent, a traditional football rivalry game, a country fair, and certain family celebrations. For all of these have those qualities I associate with a sense of place: a lively awareness of the familiar environment, a ritual repetition, a sense of fellowship based on a shared experience.

These localities are many of them out-of-date. As our cities have grown we have come closer together and acquired a more inclusive sense of community. Even so, I'm inclined to believe that the average American still associates a sense of place not so much with architecture or a monument or a designed space as with some event, some daily or weekly or seasonal occurrence which we look forward to or remember and which we share with others, and as a result the event becomes more significant than the place itself. Moreover, I believe that this has always been the common or vernacular way of recognizing the unique quality of the community we live in. The Old World farm village came to life whenever it observed the traditional farm calendar or the church calendar. The special days for plowing, for planting, for harvesting, the days set aside for honoring the local saint, were days when the local sense of place was most vivid. What made the marketplace significant was not its architecture, it was the event which took place there, the recurring day. It would be worth studying how special places have been abandoned over time, and how the event itself has been relocated.

Modern America, of course, has abandoned most of that traditional calendar. But to take its place we continue to evolve, in town after town, a complicated schedule of our own. What brings us together with people is not that we live near each other, but that we share the same timetable: the same work hours, the same religious observances, the same habits and customs. That is why we are more and more aware of time, and of the rhythm of the community. It is our sense of time, our sense of ritual, which in the long run creates our sense of place, and of community. In our urban environment which is constantly undergoing irreversible changes, a cyclical sense of time, the regular recurrence of events and celebrations, is what gives us reassurance and a sense of unity and continuity.

A remarkable book entitled *Hidden Rhythms* by Eviatar Zerubavel, published in 1981, is a pioneer treatment of what the author describes as the sociology of time: "the *sociotemporal order* which regulates the lives of *social* entities such as families, professional groups, religious communities, complex organizations, or even entire nations." Zerubavel writes that "much of our social life is temporally structured in accordance with 'mechanical time,' which is quite independent of 'the rhythm of man's organic impulses and needs.' In other words, we are increasingly detaching ourselves from 'organic and functional periodicity' which is dictated by nature, and replacing it by 'mechanical periodicity' which is dictated by the schedule, the calendar, and the clock."[1]

There is no need to dwell on the ever-increasing importance of mechanical time in modern America with our insistence on schedules, programs, timetables, and the automatic recurrence of events—not only in the workplace but in social life and celebrations. Nor need we be reminded that this reverence for the clock and the calendar has robbed

Advertisement, 1911. (From Michael O'Malley, *Keeping Watch,* New York, 1990)

much social intercourse of its spontaneity and has in fact relegated place and the sense of place to a subordinate position in our lives. Much has been written (notably by Ervin Goffman and Joshua Meyrowitz) about the disappearance of spatial distinctions and spatial characteristics because of the electronic media. In terms of the High Plains, I think it could be said that two factors contributed to an early shift from sense of place to sense of time in the organization of the landscape: the advent of the railroad with its periodicity—a decisive influence in the patterns of social and working contacts in the small railroad towns—and second, the almost total absence of topographical landmarks. Zerubavel, however, goes further in describing the social consequences of this sharing of schedules and calendars and routines, and the consequent downgrading of gathering places:

> A temporal order that is commonly shared by a social group and is unique to it [as in a work schedule or holidays or a religious calendar] to the extent that it distinguishes and separates group members from

> "outsiders" contributes to the establishment of intergroup boundaries and constitutes a powerful basis of solidarity within the group. . . . The private or public quality of any given space very often varies across time. . . . By providing some fairly rigid boundaries that segregate the private and public spheres of life from one another . . . time seems to function as a segmenting principle; it helps segregate the private and the public spheres of life from one another.[2]

So in the long run it is that recurrence of certain days, certain seasons that eventually produces those spaces and structures we now think so essential. I believe we attach too much importance to art and architecture in producing an awareness of our belonging to a city or a country, when what we actually share is a sense of time. What we commemorate is its passing; and we thus establish a more universal historical bond and develop a deeper understanding of our society. Let me quote from Paul Tillich:

> The power of space is great, and it is always active for creation and destruction. It is the basis of the desire of any group of human beings to have a place of their own, a place which gives them reality, presence, power of living, which feeds them, body and soul. This is the reason for the adoration of earth and soil, not of soil generally but of this special soil, and not of earth generally but of the divine powers connected with this special section of earth. . . . But every space is limited, and so the conflict arises between the limited space of any human group, even of mankind itself, and the unlimited claim which follows from the definition of this space. . . . Tragedy and injustice belong to the gods of space, historical fulfillment and justice belong to the God who acts in time and through time, uniting the separated spaces of his universe in love.[3]

(From Harrison Cowan, *Time and Its Measurement,* New York, 1958)

12 Looking into Automobiles

(Author's photo)

I am very pro-automobile, pro-car and pro-truck, and I can't imagine what existence would be without them. But I have learned to be discreet in my enthusiasm: disapproval from environmentalists and other right-thinking elements in the population is something I could not possibly survive. All the same, there are times when I am bewildered by the contemporary auto-oriented landscape. The younger generation takes that landscape for granted, but fifty years ago there were not only fewer cars in the streets, the ratio of trucks to passenger cars was one to eight. In 1986, three million trucks were produced in America and seven and a half million passenger cars, and from what I have recently read, the ratio is now almost one truck to every two passenger cars. I am no good at interpreting statistics, but it is obvious that the element in our work force that makes its living driving and servicing commercial vehicles is growing fast.

That is no very profound conclusion, and I imagine that some readers are thinking about how we too use our cars in our work, and in commuting, and how we expect an income tax deduction for those usages. But that is not what I mean. I mean someone who drives for pay, and the relationship between such a person and the car or truck he or she drives is much more complex and much more intimate than anything most of us know.

Pop psychology tells us that Americans cherish the car as a status symbol or sex symbol or symbol of power. That is a middle-class point of view. It suggests that most of us drive only passenger cars or sports cars. But most blue-collar Americans think of their automobile in economic terms: it is either a work tool, essential to their livelihood, or a form of capital. This is particularly true of young, low-income Americans: they acquire an old car in bad condition, or a van or a pickup, replace its engine, modify its chassis, paint it, show it off—and then sell it at a profit. Either that, or they start up a small service enterprise with it. All in all, for one who is unskilled, and poorly educated and young, there is no better way of making money than having a car. It allows him to look for a job, and when he has found it, it allows him to commute to work; and if the car is some sort of load-bearing vehicle—specifically a truck—it becomes his partner. That is why the small-time owner-operator associates his work with the world of monster trailer-tractor units with eighteen wheels, traveling overnight between Buffalo and Nashville. He shares the trucker's special attitude toward the automobile, its power, its maintenance, its appearance, its efficiency—something we laymen know little about.

Very few aspiring commercial drivers get beyond daily jobs. When I was teaching at the University of Texas at Austin with a very easy schedule and a great deal of spare time, I decided to attend a local auto mechanics school. At that time, Texas was contemplating a law requiring all full-time employed auto mechanics to be licensed, and thousands of would-be mechanics and would-be truckers enrolled in the public-supported night schools. I went to one located in an abandoned auto salesroom. There were about twenty in the class, all much younger than I. They came from a variety of backgrounds: Anglo, Spanish-American, Mexican, black—all of them blue-collar and most of them very poor. There was *one* young woman. Each student had a car, nevertheless, and many worked part-time in gas stations or garages, and one or two had trucks of their own. One young man made his living driving his ancient pickup to distribute and then collect those wooden roadside stands where Fourth of July fireworks are sold. His territory was the desert part of west Texas. All the students liked to work on their cars, even during the ten-minute breaks in the instruction, but I was surprised by how little knowledge they had of the electrical system, or what went on inside the cylinders, or even about the steering gear. I've heard it said that in Czarist Russia, when a chauffeur could not make his car start, he would beat it with a great stick. My colleagues were not so rough: when they discovered a part or a line which seemed out of order, they would gently twist it or pull on it, and take it out, and by a long process of trial and error, they usually got the motor to perform. I gathered that only a few of them had ever seen a brand-new part—even a brand-new spark plug. They either cannibalized from another car or made do with what they had. They said that they could mend *any* car with chewing gum and baling wire. I was much impressed by their ability to improvise and wondered why they had bothered to enroll in the school.

The instructor was a bright, tough young Spanish-American, an engineer who had taught this introductory course many times before, and he had no patience with any ad hoc, casual approach to auto mechanics. When he saw a specimen of it, he ripped it out. "Do it the right way," he yelled, "read your manual, use your head! Be correct!" I doubt if the trainees knew what the word meant. Nevertheless, they reluctantly and slowly did the job over. They realized that they would soon have to take an exam, and pass it, if they expected ever to get a job.

The final exam was hard. Each trainee had to give a complete public tune-up and lubrication job to one of the other trainees' cars chosen at random by the instructor, and to explain every part, every tool, every procedure to the class. Most of them passed, and then went on to take a second, more advanced course in electronics and the use of diagnostic equipment.

That was when I dropped out, but I too had learned something which I have tried not to forget: if you love and respect your automobile the way these young men did, and if you depend on it for your livelihood, the automobile will reciprocate, as it were, and teach you many useful things: it can teach you to be accurate, teach you to use the right tools, teach you to make decisions. It cannot teach you the difference between good and evil, but within a somewhat restricted realm it can teach you the difference between right and wrong: between correctness and sloppy, dishonest work. In a word, learning to be a good auto mechanic is learning to be civilized. The trainees had not the chance to learn those lessons anywhere else: none had gone to high school, and they had all left home. Their tough young teacher had bullied them into a kind of maturity, but I think it was their love of automobiles which accounted for their transformation. They wanted to be a part of the technological world around them.

No doubt many of the trainees later relapsed into their old slovenly ways of working and thinking. Even so, I now look at certain features of the auto-oriented world with more respect. Whenever I see men anxiously peering under the raised hood of a beat-up car parked in the front yard of their house, I imagine that they are disciplining not only their hands and their eyes, but also their minds, and visualizing and trying to re-create a kind of order. I'm sorry to see the junk they scatter all over the place, but they see some value in it which escapes me; a means of making a living. When Dr. Johnson had the job of auctioning off the brewery of his friend Piozzi, he said, "We are not here to sell a parcel of boilers and vats, but the potentiality of growing rich beyond the dreams of avarice." Remember that when you pass an auto junkyard.

13 Truck City

The twentieth-century reality: architecture in second place, heavy truck traffic dominating. (From Mick Harner, *Transport*, New York, 1982)

Today there are more than 180 million cars in the United States. In 1908 there were less than 200,000. Two years later there were close to half a million. For the next few years the number increased annually by more than a third.

From the beginning Americans were fascinated by the automobile, though few families could afford to buy one; when the average worker earned a dollar a day, a medium-priced automobile, costing more than a thousand dollars, was clearly only for the rich. Yet the thought of how the automobile was going to transform and enrich American life was stimulating, and we followed every step in its evolution. We discussed speed records, endurance records; we applauded improvements in the motor, and the daredevil feats of drivers and racers. We were especially interested in elegance of design, and what we most admired was the passenger car, also known as the touring or pleasure car. The rich and sporting element in American society adopted the automobile because of the mobile and adventurous way of life it fostered. The social pages gave details of the groups of fans, the meets and rallies of exclusive automobile clubs, and told of who had attended the *concours d'élégance*. Academics and admirers of the traditional rural culture sought to turn us away from the automobile: in 1906 Woodrow Wilson declared that the automobile represented an ostentatious display of wealth, threatening to incite envy and class feeling; but the opposite proved true: we loved the passenger car and dreamt of when we too could own one.

So much favorable publicity had the effect of rousing general interest in the manufacture and sale of automobiles. Investors and manufacturers and engineers soon took the new invention seriously and promoted research and experimentation. Yet the greatest appeal lay in the kind of lifestyle it promised: the dawning century was to see a new culture: emancipated, healthy, infinitely mobile and promising hitherto unknown pleasures and experiences—all because of the automobile.

The commercial automobile, however, the bus, the taxi, and especially the truck, were all excluded from this cordial welcome. In 1910 there were only eight thousand trucks in the country, and most were so uncouth in appearance, so disappointing in performance that they belonged in a special category. In the years before World War I it was the custom among the drivers of passenger cars to show their solidarity by tipping their hats to one another. (A similar custom prevailed until recently among motocyclists.) But the drivers of trucks and vans were never included in this exchange of high signs. Given the social implications of the passenger car, it is easy to understand the discrimination, for the passenger car stood for a leisurely suburban or rural existence, based on family togetherness, love of nature, and a good income. The truck, on the other hand, stood for work, and work of a menial kind.

The word *truck* had been used in colonial times to describe a heavy wagon capable of carrying a load. With the coming of the automobile, we redefined the word to describe an *automotive* vehicle designed to carry a load, and that is still the accepted definition. It is a poor one, nevertheless, for it is used to include the pickup, the van, the minivan, and even the jeep; these are all supposed to be trucks. But in fact, in everyday parlance we almost always use the word *truck* to mean a large vehicle which can haul heavy loads for some distance, and does so *for pay*. Vernacular usage often defines an object—whether a house or a car or a piece of furniture—not by how it is *made* but by how it is *used*, and most trucks are used to make money. To be sure there are marginal cases: the pickup, like the jeep, is now undergoing a process of suburban gentrification: both are popular in leisure activities and in certain off-beat sports. At one time the station wagon was a modest passenger car equipped with extra seats and was called a depot van. It was a money-making vehicle then, but look at it now!

Many of those eight thousand trucks in 1910 were small electric delivery trucks, used by such businesses as laundries and bakeries and dairies, and by the U.S. Post Office, for making daily deliveries along established city routes. They were a great improvement over the horse-drawn wagons or surreys or buggies previously used. They made no noise, they did not smell, they were easy to drive, and very sturdy. In appearance they resembled the horse-drawn wagon, and in passenger cars this last feature appealed (so we are told) to that die-hard element which was still pro-horse and anti-automobile.

But the electric van or truck had a serious shortcoming: it ran on batteries (large and heavy, slung under the chassis) which had to be frequently recharged. At best they could go eighty miles without having to turn in at one of the infrequent recharge stations, and the operation took time and cost fifteen dollars. Largely because of the weight of the batteries, the electric truck had trouble climbing hills. Finally, it was expensive. No working man could think of buying one and going into the delivery business on his own. But would he have wanted to? Driving an electric delivery van for pay, even if only a dollar a day, probably had its appeal. It was safe, relaxing, and without surprises. But Americans even in those days had a distinct impression that the automobile represented freedom and personal responsibility. A vehicle like the electric van which had to stay in the city and abide by a fixed routine was in a class with the railroad train or the streetcar: confined to a track and a timetable. In any case, as gas trucks and passenger cars became cheaper and more efficient, the electric car dropped out of favor. It lingered until 1925 as a sedan, small and elegant and eminently suited to lady drivers. Then it too vanished.

There remained that other contingent out of the original eight thousand trucks. They were gasoline-powered, and in fact the gasoline truck had been around as long as the gasoline passenger car. But it had failed to show early promise. For one thing, it had been given none of that encouragement lavished on the passenger car by social and financial leaders. No one attended the strenuous truck endurance runs, and the press had little to say about the experimental models being produced by amateurs: inventors, mechanics, retired bicycle makers, dedicated tinkerers working in stables and blacksmith shops, or on the road. A few manufacturers of trucks produced a dozen or more examples, then went bankrupt or were bought out. There were no standard parts, only parts taken from carriages or pieces of machinery. Trucks often unhappily resembled in appearance the conventional wagon—wheels, dashboard and all—nor did they always perform much better, for they were slow and heavy, far heavier than the load they carried. Steel-rimmed wheels gave little traction on the smooth city streets, and motors deficient in power prevented them from traveling on rough and muddy country roads. It was generally agreed that all these hybrid vehicles could be used for was hauling freight from the railroad station to the factory or the warehouse, or making routine deliveries, like the electric van, within the city.

An electric truck from the 1920s.

Trucks in those early years were usually owned by shipping or distributing firms, or by such public agencies as police and fire departments. They were too expensive, too unreliable for the average small business. They operated in fleets: each morning would-be truck drivers went to the yard where the motley collection of trucks was parked and waited to be given a job for the day. At that period most drivers knew nothing about maintaining and repairing a truck, especially if it was one they had never driven before, and often they returned their vehicle at the end of the day in bad condition. Strictly speaking, the truck driver was an unskilled day laborer, easily replaced. He was assigned a clumsy, unfamiliar tool to do a job which for him had no future, and his status was not much higher than that of the factory hand or field hand. He was paid two dollars for a twelve-hour day.

The trucking industry might well have continued in that manner: with fleets of trucks operating like the old fleets of taxis. But two surprising developments occurred. First, the truck became a versatile and adaptable piece of mechanical equipment, capable of doing much more than simply hauling heavy loads and coming back empty. Second, it broke away from centralized control and in time provided many men with small businesses of their own. It thus became an important part of an American vernacular way of life, and remains so to this day.

In these transformations the manufacturers at first played a minor role. Most of the experimentation came from small-time mechanics and engineers, and it was they who in effect created a whole new industry—not a cottage industry but a garage industry; it was largely they who produced the modern truck.

Many were farmers, for farmers were acquainted with stationary engines long before they had seen automobiles. They were part-time mechanics, electricians, tinsmiths, blacksmiths, and had already automated much of their work. It was natural for them to experiment and find out how this new machine could be used for other purposes. One problem was getting their vegetables and dairy products promptly to city markets. Railroad schedules were often inconvenient and involved time-consuming transfers and delays. It was a master blacksmith in a small Wisconsin farm town who first produced the four-wheel drive, which enabled trucks to negotiate bad roads and cut across fields, pick up the produce, and take it directly to the distributor. It was another small-town mechanic who built the first auto trailer, which often doubled the truck's capacity. In 1915 pneumatic tires were introduced, and another now forgotten innovator gave the truck headlights, enabling it to travel by night; others devised the dump truck and the truck with a low bed for loading heavy loads and for dockside loading. None of these were technological breakthroughs, but they contributed to a redefinition of the truck, and eventually to its being used for more than simply transportation. Its increasing versatility made it essential in heavy construction, in road building, and in the extraction of hitherto inaccessible natural resources—in oil fields and forests located in rough terrain where the railroad could not go.

The Ford Model T, introduced in 1908, proved to be particularly useful on the farm. Though strictly speaking it was a small passenger car, any ingenious owner could transform it into a pickup truck for carrying produce or even transporting livestock. Once a pulley was attached to the crankshaft or rear axle, the Model T provided power for sawing wood, pumping water, grinding feed, or furnishing electric light in emergencies. In the 1920s several farm equipment firms offered attach-

An improvised truck. (From *Fortune,* November 1933)

ments designed to turn the Model T into a cultivator or tractor; but these proved impractical.

There is undoubtedly an axiom to the effect that any tool which has proved efficient in accomplishing its original function is to be used for some other, totally unrelated job—and used successfully. That would account for the truck, even before it was twenty years old, being used as a snow plow, as a digger of post holes, as a road grader, as a temporary platform for derricks, and as a source of emergency power. More important, it was also used not simply to *carry* a load but to *process* or modify the load, even as it traveled on the highway. The mobile cement mixer, familiar to us all, appeared as early as 1916, and the refrigerator truck (controlling the temperature of its load) appeared in 1920. The modern ambulance is actually a truck equipped to give medical aid to its passenger. The evolution of the truck has not come to an end; and while the passenger car, no matter how luxurious, still remains faithful to its original function of carrying passengers, the truck has in many instances become a part-time office, or a part-time workshop, part-time operations center, and part-time sleeping quarters. The introduction of the two-way radio and car phone has given the truck and its drive a greater freedom of movement and of decision making; so much so that it can be said of certain trucks that they are actually small mobile business or service establishments; and a large element in urban traffic consists of such mobile workplaces.

By the 1930s the truck had become a much more sophisticated vehicle: larger, more reliable, more powerful, and more sightly. Even though trucks constituted only a minority of automobiles, they had become increasingly essential in many types of business enterprises, largely because they could accommodate small loads at short notice and could go directly to the customer. Their final acceptance as a form of commercial transportation—for short distances in partnership with railroads—came in the late 1930s.

At that period the typical American factory was likely to be a multistoried brick building with many windows located in the built-up downtown section, as near as possible to the railroad. At that time that was the best place to be; for transportation for both raw materials and the finished product, access to local supplies of coal. The labor force, living nearby, often came to work on foot. Multistory construction, caused by lack of space, allowed for gravity flow in the manufacturing process, and usually allowed for the use of natural light—hence the many windows.

But by the 1920s there were reasons for changing the layout. The automobiles of commuting office workers crowded the narrow downtown streets and made shipping and receiving a nightmare for factories. New machinery called for sturdier construction; yet there was no room for expansion. No doubt the most compelling reason for wanting change were the new theories of industrial organization propounded by F. W. Taylor and his followers. Though primarily meant to introduce more efficient work methods, Taylor's "scientific management" involved a radical reorganization of the industrial plant itself. It proposed a factory layout emphasizing continuous, uninterrupted *horizontal* flow in the whole production process, a speedy in-and-out of raw materials and finished products, and a great reduction in storage and space for storage. The assembly line, first introduced by Ford in 1913, was the logical product of Taylor's management methods.

Scientific management proved to be a popular idea, and it was not long before new industries and the older and larger ones started to move out of the congested city into the outskirts where space was available.

Remote from the built-up city streets and from the railroad lines, there soon emerged many examples of the kind of factory and warehouse we are now familiar with: one-story constructions, often windowless, with vast interior spaces interrupted by few reinforced-concrete or steel columns or uprights—where a variety of mobile conveyors and forklifts and miniature industrial tractors can circulate freely. The light, whether from skylights or artificial, is diffused and the pattern of movement is everywhere horizontal. Unlike the situation in the center of the city, here in the open landscape the factories are not close together. Surrounded by their own spacious territory, they sit at some distance from one another, many with well-kept lawns and trees. With no water tower, no smoking chimneys, only a discreet sign next to an imposing entrance driveway, they often have an institutional appearance, as if they were centers of research.

Here the truck comes into the picture: not one truck, but trucks by the dozen—in some cases large trailer-tractor units capable of hauling thirty tons or more—immaculate, identical, are backed up in a row against the long loading dock.

The loading dock is a feature of the modern factory not much discussed in architectural circles, but it is not only an essential part of the building, it has to be designed and built with great precision. Its level and the level of the truck waiting to be loaded (or unloaded) have to be the same; the continuous horizontal movement of the interior has to be extended by means of forklifts into the truck itself. The loading dock in turn relates to the vast space surrounding the plant devoted to the parking and maneuvering of the trucks, as well as to the parking of the employees' cars. Finally, the whole complex leads out to the highway. Loading docks, along with extensive parking areas, are something new in factory layouts, and what has evolved is an industrial structure with two facades: one for the offices and the formal entrance, the other for shipping and unloading. This obvious arrangement has now become general in all establishments depending on commercial vehicle traffic: the dark and rubbish-littered rear area in the factory complex is a thing of the past.

The road which the flow of trucks follows until it reaches the interstate passes through an empty and spacious landscape of isolated factories, truck terminals, warehouses with a scattering of road-oriented service stations, trailers, landfills, and what were once fields: a landscape

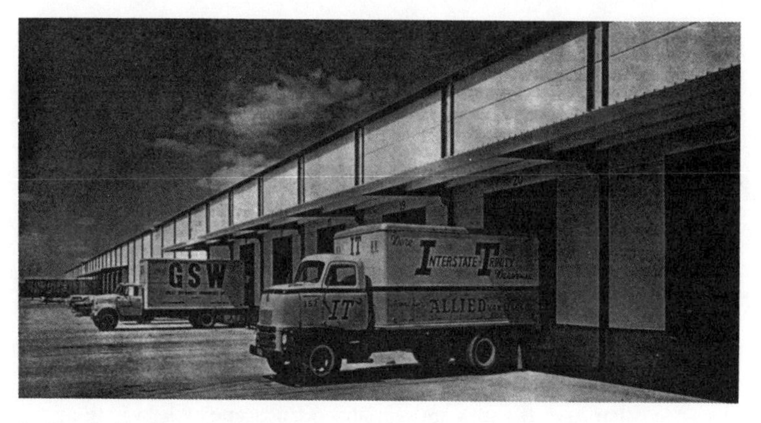

Loading dock with trucks. (From *Urban Land Institute Bulletin* 44, Washington, D.C.)

waiting not to be lived in but to be built on; deserted after work hours and dominated by the curving pattern of roads and projected roads, power lines, and billboards saying "12.75 acres available, zoned C"; with not a pedestrian in sight. We have all passed through this landscape on our way to the airport or to the interstate.

Have we forgotten how we responded to that landscape when it was only beginning to take form a half-century ago? In those years after World War II, we were on the lookout for a new world of peace and prosperity and leisure, and that new landscape contained many elements which we thought held promise. That was when the general public first became aware of what we then called "modernistic architecture"—the International Style; and those one-story factories stripped of ornament, functional in layout, helped us understand and accept the new idiom. Commenting on the modern horizontal industrial factory, John B. Rae wrote that "the modern warehouse is no longer a store house but a 'transit shed' for continuous inventory replenishment. More space is devoted to processing orders, docks, and aisles for self-propelled vehicles than to actual storage."[1] That absence of storage space and the emphasis on flow was typical of the factory as well, but the average American was already familiar with other versions of the transit shed: the supermarket, the drive-in restaurant, the great variety of businesses from the liquor store to the bank where customers as well as goods and services were part of the environment organized for flow and the elimination of storage or time-consuming transactions. Instinctively we discerned in the new landscape and its buildings aspects of a bright future.

We marveled at the new roads and highways, anti-urban in their green, semipastoral margins and in their rejection of the grid. The large-scale sweep of the new interstates, with their cloverleafs and overpasses and their steady, uninterrupted flow, reminded us of the visionary displays of the road of the future in the Futurama of the 1939 World's Fair: bypassing the city, heading out into the green countryside where housing developments and trailer courts spoke of young families starting out in life.

What gave the new industrialized landscape style was the presence—in parking lots, lined up at loading docks, barreling along the new white highways—of trucks: enormous and sleek and shiny. Since we ourselves had cars and went to work in cars, we felt an affinity with the burgeoning truck culture of truck stops identified with home-style southern food and truckers' songs in the juke box; for the trucker was replacing the cowboy as folk hero. Watching the evolution of the drive-in, driving through the highway-dominated landscape with its new spaces, its brightly colored signs and structures, seemed a good way of observing our progress toward a new social order.

It cannot be said that this response represented a conscious rejection of the traditional landscape order; but it is not totally remote from the visions of the future propagated by the design professions. In the popular vision of the new city there were elements—distorted and oversimplified—of Ebenezer Howard's Garden City, of Le Corbusier, even of Lewis Mumford; expanses of greenery, small clusters of dwellings, landscaped highways, and a feeling for new architectural forms, for clarity and accessibility and freedom to move.

But that was a half-century ago, and neither the discovery of the new landscape nor the landscape itself has survived. The elegance of its new architectural forms is now concealed or modified, the open land of the earlier years has been subdivided, and even the splendid space of the new highway becomes increasingly congested; uninterrupted flow means more often than not uninterrupted traffic jams. The old factories and warehouses are rediscovered, remodeled, gentrified, while the new-style factories hide as best they can behind a miniature wilderness. The truck, finally, is identified with noise and pollution.

The ultimate transit shed: a drive-through bank. (Photo: Tom Strong)

What is left? A vernacular, blue-collar version of the truck-oriented landscape of the urban fringe has moved into the city and is destroying the traditional urban culture. When we venture beyond the traditional center of town into the less prosperous neighborhoods, we see that the transit shed in several versions is by way of becoming the standard commercial building, and even the standard institutional building. It is a form suited to enterprises operating with small inventories—chain stores, gas stations, discount houses, art galleries, even museums and libraries: all depend on services and supplies brought in at regular intervals. Even the prestigious streets of the city, with their hotels, restaurants, department stores, and office buildings, operate on the same Taylor-inspired principles of steady in-and-out flow, horizontality, and the elimination of storage. Main Street, which not so many years ago was still lined by impressive, more or less uniform facades of conventional architecture and still retained an atmosphere of permanence and limited accessibility, is now perforated by drive-in entrances, parking lots, underground garages, service alleys, and the brief appearance of transient business enterprises.

For the generation which still remembers the traditional street and its monumentality, this transformation is a depressing spectacle, and it does not help us to be told that in the past cities have frequently suffered the same decay of the central area. In many cases they have survived by moving the central city to another location. In Europe popular wisdom says that cities move to the west, that the center of power and prestige always shifts in that direction.

Were those cities the victims of a new kind of traffic in their streets? By traffic we mean not simply the vehicles which take over the street and threaten to eliminate existing forms of transportation, but also people and loads: those are the elements which bring about change. The Trojan horse was welcomed by the inhabitants of Troy; it was when that wooden horse disgorged its passengers that there was dismay.

In the case of the American town or city, the automobile—especially the commercial automobile, the truck, the pickup, the van and minivan and jeep—has been most effective in introducing a different spacial order. For what those vehicles contain (and distribute) is not only new attitudes toward work, new uses of time and space, new and more direct contacts with customers and consumers, but new techniques of problem solving. One reason factories and other industrial enterprises have tended to congregate in special areas is that they locate where they can make use of what are called "external economies"—specialists and subcontractors who can take on short-term jobs. The mobile specialist can in many cases provide that help, no matter where the client is located, and in consequence the traditional concentration in the downtown area of specialized crafts is no longer essential: even the shopping center and the mall can serve as go-betweens uniting consumer and specialist, thanks to mobile collecting and delivery and service operations.

But aside from its impact on the central city, the current small-scale commercial vehicle is introducing into the newer and low-income sections of the city a kind of vitality and movement which seemed to have vanished. The working-class dwelling, already reduced in size, is fast losing its self-sufficiency and depends more and more on mobile services. The commercial street, bordered by one transit shed after another—supermarkets, used-car lots, fast-food outlets—is in itself an elongated transit shed devoted to steady flow automatically controlled by traffic lights; and the traffic is largely composed of mobile enterprises—delivering, collecting, hauling, and distributing within the neighborhood. Social interaction moves into the street and takes the form of cruising or gathering in parking lots or around gas stations.

As of now the proportion of trucks—vans, pickups, jeeps, and medium-size trucks for city use—to passenger cars is approaching one in two. The operation and servicing of the growing number of trucks constitutes a respectable sector of our economy. Nor are these vehicles confined to the business sector. If the old establishment residential streets seek to exclude commercial traffic, there are vast areas of the city where small-scale commercial and service traffic is welcome. What keeps those trucks and vans and pickups and converted passenger cars contantly on the move? They are hauling small loads, sometimes hauling passengers, but they are also making repairs, installing and removing and replacing and servicing small businesses. Before the proliferation of these vehicles, remoter areas of the city suffered from neglect and social isolation. Now they are once more related to the city, and even the poorest of households is, at least in theory, within reach of help. Ascribe that not to a zealous welfare department or an efficient public transit system or a new clustering to foster a sense of community, but to the presence of a highly mobile sector of artisans and craftsmen, mechanics, maintenance personnel, small-time franchise holders who will come to any customer and bring reassurance that he or she has not been forgotten.

The small commercial or service vehicle is helping weave together the city which an earlier generation of automobiles had torn apart. Both kinds of vehicles, the passenger car and the truck, share responsibility for decentralization, but each has begun to mitigate the effect of the other. The passenger car has relocated many experiences and pleasures (once identified with private life) back in the public realm and the neighborhood; the truck has reintroduced small-scale services and skills into the private realm and new communities. Both responses are helping create a vernacular type of city; loosely structured, fluid, and expansive.

14 Roads Belong in the Landscape

(From Drake Hokanson, *The Lincoln Highway,* Iowa City, 1988)

Which came first, the house or the road leading to the house? Medieval scholars with their love for origins and symbols may well have long wrestled with the question, eventually coming up with a theological counterquestion: Which of the two objects had been divinely ordained to *be* first? It could have been reasoned that if God had meant us to stay home, to be sedentary, to put down roots as farmers or husbands (a word which once signified house-dwellers), he would have first commanded us to build a house. But if he had intended us to be forever on the move—hunters or herders or pilgrims in search of an elusive goal—he would have ordered us to beat a path, to make a road and follow it.

In medieval times, the past was seen as a series of migrations and invasions, of endless wanderings over the face of the earth. But in later, more settled centuries, the above question was interpreted in more mundane terms: the one coming first would not necessarily have been the earliest but the one conferring power and prestige; and from that point of view "house" was the right answer. House is much more than shelter. It implies a territory, a small sovereignty with its own laws and customs, its own history, its own jealously guarded boundaries. House stands for family, for dynasty. However modest it may be, it still has its place in the elaborate spatial hierarchy of the European world: kingdoms, principalities, domains—then the house.

Compared with these attributes—still venerated—what could the road offer by way of qualifications for first place? Certainly it played an important role in our daily comings and goings: it might even be said that it was the road which first brought us together in a group or society. Yet the purpose of every road or lane or path is to lead to a destination, and the question itself presupposes a house. So the true function of the road is to serve us by taking us home. Without a specific destination, a road has no reason for existing. Left to its own devices it tends to wander into the wider environment and disappear. It has another tendency, much more dangerous: to introduce unwanted outsiders into the self-sufficient community or house. Finally, theology again enters the picture. The first man to hit the road was Cain, the murderer of his brother, and cursed to be a fugitive and a vagabond; Cain, the first man to build a city.

Disqualified by its own genealogy, outclassed by the prestige of private space, the road has long suffered from neglect by historians and students of the landscape: dismissed as an unsightly, elongated, crooked space used by merchants and ravaging armies and highway robbers, whereas the house (as we learn from Joseph Rykwert's essay *Adam's House in Paradise*[1]) has become the symbol of arcadian simplicity and innocence.

That simple question, at least in its cryptic form, is one we no longer bother to answer. We are now less interested in origins than in what comes *after:* specifically, the relationship over time between these two familiar features of the landscape. The relationship has never been easy. We are just emerging from a centuries-long period when the road was subservient to place and given little respect. Today, a hundred years after the invention of the automobile, the question would be answered in favor of the road—or its modern version, the highway—which continues to weave a tight, intricate web over every landscape in the Western world and has spawned a whole breed of roadlike spaces—railroad lines, pipelines, power lines, flight lines, assembly lines. Now the question demands a very different sort of answer: Which do we value more, a sense of place or a sense of freedom? What clouds the debate is the insistence by both parties on defining the road as a disturber of the peace, an instigator of radical change: welcome change in the eyes of the land developer and traffic engineer and urban reformer; in the eyes of the environmentalist and history buff, change as the destroyer of privacy and the acceptable status quo.

The answer will come when we define or redefine the road as it exists in the contemporary world; when we recognize that roads and streets and alleys and trails can no longer be identified solely with movement from one place to another. Increasingly they are the scene of work and leisure and social intercourse and excitement. Indeed, they have often become for many the last resort for privacy and solitude and contact with nature. Roads no longer merely lead to places; they *are* places. And as always they serve two important roles: as promoters of growth and dispersion, and as magnets around which new kinds of development can cluster. In the modern landscape, no other space has been so versatile.

Odology is the science or study of roads or journeys and, by extension, the study of streets and superhighways and trails and paths, how they are used, where they lead, and how they come into existence. Odology is part geography, part planning, and part engineering—engineering as in construction, and unhappily as in social engineering as well. That is why the discipline has a brilliant future. When archaeologists uncover a pre-Columbian road system in a Central American jungle and speculate about its economic or military origin, when geographers study the environmental impact of a new highway in an isolated region, and when a city council decides to change a two-way street into a one-way street, they are all thinking in odological terms: in terms of the function of the road in question, in terms of its impact on the landscape, and in terms of traffic.

They are thus thinking in a strictly modern manner. It is only within the last two centuries that we have recognized the wisdom of building roads to last and of building them to serve a specific type of traffic. It was when commercial vehicular traffic increased (as a consequence of industrialization) that we gave thought to their location and subsequent maintenance. In the eighteenth century something like a science of road and highway engineering finally emerged.

But a further stage in the evolution of odological techniques came after World War II, when an immense increase in heavy automobile traffic took place throughout the world. The old nineteenth-century definition of road—"a passage between two places, wide and level enough to accommodate vehicles as well as horses and persons travelling on foot"—proved inadequate. We now see the highway as an essential element in the national or regional infrastructure, and the solution to problems of congestion and surface deterioration is no longer the building of more (increasingly costly) highways but the controlled and efficient use of those we have.

Consequently, contemporary odology involves much more than improved methods of construction and maintenance; it devises new traffic control systems, separate networks for different kinds of traffic, greater control over roadside and road-oriented development, and higher user fees; and we may soon reach the point in metropolitan areas of giving priority in the design and location of roads to odological rather than environmental factors.

We do not always give credit to how the motorized American—commuter, tourist, truck driver—has accepted the new odology, how docile we have been in complying with the scientific definition of the highway as a managed authoritarian system of steady, uninterrupted flow for economic benefits. Within a few decades we have learned to abandon our traditional attitudes toward the road and to adopt new driving skills, new ways of coping with traffic, a whole new code of highway conduct and highway law; learned to accept without questioning a vast and growing assortment of edicts indicated by signs and lights and symbols and inscriptions on the surface of the road itself. We have learned to drive defensively and to outwit traffic jams and lurking policemen. We have also learned to take advantage of the proliferation of highway-oriented businesses and diversions and to discover the joys of speeding, of seeing the landscape flash by at an inhuman rate. We have become so submissive that radical odologists are encouraged to propose further electronic controls within our own vehicles, further restrictions on our use of the highway, further tolls and fees.

That is the price we pay for uninterrupted steady flow, and no doubt the price is reasonable. But odologists seem to forget—and we ourselves sometimes forget—that the road serves other needs. For untold thousands of years we traveled on foot over rough paths and dangerously unpredictable roads, not simply as peddlers or commuters or tourists, but as men and women for whom the path and road stood for some intense experience: freedom, new human relationships, a new awarenes of the landscape. The road offered a journey into the unknown that could end up allowing us to discover who we were and where we belonged.

The superhighway has an essential role to play in our workaday lives; even during the 1992 Los Angeles riots the red traffic lights were respected: they represented an order transcending the political or economic order. But we must reformulate the current technological definition of odology so that it recognizes the *private* experience. There is still time.

Edgar Anderson was an eminent botanist best known for his investigation of the origin of many New World cultivated plants and of their diffusion throughout preconquest America. In 1967 he published *Plants, Man, and Life,* an informally written account of his theories and of his botanical travels in Mexico and Guatemala. It told a wide public something of the complex story of the domestication of plants, and of how our own landscape had been transformed by the migration of the many wild as well as cultivated plants from Europe, though their original home in most cases had been Central Asia, thousands of years ago.

> Take the autumnal aspect of the central and eastern United States, roughly the region from Boston and Philadelphia to Minneapolis and Kansas City. In all that area, green in the autumnal landscape is a measure of European influence. The green grasses of pasture and roadside, the green trees of orchards and parks, are greens which have come with us from Europe. Our own native flora is bred for our violent American climate. It goes into the winter condition with a bang. The leaves wither rapidly, they drop off in a short time, frost or no frost. . . . European trees and grasses color slightly and slowly if at all; our native grasses are as bright with color as our native trees. . . . If you want to know how much of the landscape in which you spend your days is authentically American, look around your hometown in mid-autumn. As man moves about the earth, consciously or unconsciously, he takes his own landscape with him.[2]

It was his unflagging curiosity about that everyday landscape of cornfields and backyards, the margins of roads and streams, the abandoned fields of the countryside that Anderson communicated to his readers and to his students. He later told of how he took his botany classes out to investigate the plant life in "dump heaps and alleys. . . . We study Trees of Heaven, weed-sunflowers in the railroad yards, wild lettuce on a vacant lot. Gradually a few of us are beginning to accept man in our own biological limitations as a real part of nature. The ecology of dump heaps should be more rewarding for the time spent than the ecology of grasslands of the Great Plains. . . . In the dump *homo sapiens* is the most overwhelming of all the organisms in his primary and secondary effects on the landscape under analysis."[3]

Anderson was a special kind of botanist: an ethnobotanist, one who studies the relationship between plants and human beings; and what ethnobotany tells us about man's very earliest experiments in agriculture offers a new approach to nature. That America has become interested in the green nature in cities and farms can be ascribed largely to Anderson's writings and lectures.

Landscape magazine, which I started publishing in 1951, owed him a special debt of gratitude. When it was still a small publication, uncertain as to its message and almost completely unknown, Anderson generously offered to write occasional brief essays for us. He became our first contributor of note. He also provided the magazine with something like a policy, a point of view: a humanized version of the current concern for the natural environment, a sort of "ethno-environmentalism," dwelling on the exploitation and discovery of the ordinary human landscape of America, including all the modifications and changes and destruction produced by man. Anderson, indeed, encouraged a critical examination of environmental fanaticism and expressed strong disapproval of "the amateur Thoreaus and the professional naturalists of our culture [who] have in the United States raised the appreciation of nature to a mass phenomenon, almost to a mass religion; yet at the same time have refused to accept man as part of nature. . . . They are the chief ultimate sources of our unwritten axiom, that cities are something to flee from, that the harmonious interaction of man and other organisms can only be achieved out in the country."[4]

One of the first essays published by *Landscape* was a nostalgic evocation of the "horse and buggy" landscape of Anderson's youth: a description of the sights and sounds once part of slow travel through the rural countryside at a period when the narrow, ungraded roads were little more than paths through fields and woodlots. In fact, much of Anderson's investigation of the botany of disturbed areas concentrated on roadside vegetation. A favorite plant of his was the common sunflower. It had several attractive characteristics: it was one of those organisms which Anderson called "camp followers": plants and animals, not always domesticated, which choose for very good reasons to live near people and to follow them when they move elsewhere: weeds and insects, rats and mice (and now we can include raccoons and deer and a variety of birds), as well as less obtrusive plants like daisies, yarrow, buttercups, and plantain. All of these, and many more, have long associated themselves with human habitats, beginning in the remote Neolithic time when we first began to disturb the surface of the earth.

The second appealing characteristic of the sunflower is that it is the only plant domesticated in pre-Columbian times in what is now the United States. This means that from the ethnobotanical point of view it is a very suitable plant to study. Given sufficient botanical, anthropological, and historical data, we can find out when and where the sunflower was first domesticated, and how it later became a camp follower: a plant with a preference for full sunlight, and for terrain altered or disturbed—usually (though not always) by man.

"Why should this be?" Anderson asked himself. "What is there about the presence of man that stimulated his plant and animal companions into increased evolutionary activity?" Anderson suggests that sunflowers, as well as many other camp follower plants, dislike grass or turf. "One thing they seem unable to tolerate is competition from other kinds of plants, particularly grass. . . . It is not that sunflowers do not appreciate rich soil. . . . They are on the poorest soil because only there can they get away from grass."[5] The most likely explanation is that sunflowers evolved this preference within historic times; when Neolithic man began to dig and burn and plow, he incidentally created a new micro-environment: an environment where the soil was open and bare and from which all existent plant life had been suddenly removed. That, of course, is true of every plowed field; but this preference on the part of many camp follower plants for open, bare spaces, and their absence from the neighboring meadow or rangeland, are a vivid illustration of how man has not only destroyed ecosystems but also devised *new* ones.

What were the consequences of such human intervention? Primarily it allowed for the development or evolution of other plant species. "The native vegetation [in undisturbed sites]," Anderson told the Conference on Man's Role in Changing the Face of the Earth in 1957,

> has a long evolution of mutual adaptation. Plants and animals have gradually been selected which are adapted to life with each other, like pieces of a multitudinous jigsaw puzzle. It is only when man, or some other disruptive agent, upsets this whole puzzle that there is any place where something new can fit in. If [in the original undisturbed site] a radi-

> cal variant arises, it is shouldered out of the way before it reaches maturity. In a radically new environment, however, there may be a chance for something new to succeed. Furthermore, the hybrids and their mongrel descendants were not only something new. They varied greatly among themselves. If one of them would not fit into the strage new habitat, another might.[6]

Not every plant (nor every animal or insect) will adjust itself to such open habitats as dumps or highway cuts or bombed-out cities—or even to well-manicured flowerbeds. "The average plant," Anderson remarks in discussing certain attempts to grow ornamental flowers on roadsides, "is awfully finicky about just where and when it will grow. . . . Precise limits of temperature and moisture have been worked on for some plants but the vastly intricate business of which plants they will or will not tolerate as neighbors, and under what conditions, has never been looked into except in a preliminary way for a few species."[7]

The apparently random concentration of certain plants in areas of disturbed soil offers us a valuable guide in any program of managing or transforming the vegetable cover of a new environment; and Anderson in all his writings has underscored one message: Man is a maker of new plants and new plant communities. "The detailed study of this process should illuminate for us the course of evolution in prehuman times . . . and most importantly, should enable us at last to understand and eventually to control the living world around us."[8] It is when we recognize the role we have played and continue to play whenever we plow a field, put in a garden, preserve an endangered species, or build a road that we learn a greater awareness of our relationship to the green environment.

But in fact, long before man appeared on the scene, the face of the habitable earth was everywhere scored and crisscrossed by the paths and tracks and trails made by animals. Some led to waterholes or salt licks or expanses of edible grass. Some were lengthy routes of annual migrations, and others were made by animals in search of other animals to kill and eat. Whether they were in the forest or the grasslands or in the desert, all the paths were bare of vegetation, and smooth. They avoided obstacles and uncertain terrain by taking a meandering course.

Some two million years ago an apelike hominid began to come down out of his habitat, the trees of the forest, and to take tentative steps in the open grassland and use the existing paths. He was small and inconspicuous, and survived by foraging; but the way in which he moved and his reaction to the presence of other creatures marked him as distinctively different.

In the past, strange changes in his apelike anatomy had started to appear. His long, dangling arms became shorter, his hands developed a powerful and versatile thumb. Whereas the feet of apes were flat, not suited to a prolonged erect posture, the feet of this hominid acquired an arch which gave his manner of standing and walking a greater balance and self-assurance. Most important, his eyes, formerly focused on what was on either side, now focused on what was ahead, and were capable of stereovision, of seeing a wide segment of the world surrounding him.

He eventually became a skillful maker of tools and an effective hunter. He developed needs unlike those of other animals: the need for certain stones for tools or weapons, for certain special kinds of wood; he developed the need to store and conceal certain objects for future use—all of which meant that he undertook to make his own paths, often into remote areas. Eventually his familiar environment was marked by a number of specialized spots: those associated with a resource, an event, or a memory. It might be said that he developed a sense of territory, a protolandscape to protect.

Our generation has seen a remarkable growth of scientific interest in the mental evolution of early man. Though there is still much to learn about his anatomical evolution and his adaptation to the natural environment, more and more we are asking about his intellectual and spiritual unfolding: When and how did he acquire a mathematical awareness, an awareness that time and space could be measured? When did he first develop a taste for ornament? When did he become an artist? These questions we are now asking of the psychologist, the philosopher, the art critic, the sociologist, and the answers are coming in.

Has odology any help to offer? Most animals rely on their sense of smell to guide them along paths, and to identify others of their own kind, and their territory. But early man and we, his successors, recognize people and classify them by their appearance, and use our sense of vision in finding our way. What we register in passersby are their posture, their gait, and how they respond to the presence of others. Since this kind of contact with strangers usually occurs in a public area, the road or path becomes the first and most basic public space. Moreover, for each of us it becomes important as the place where we in turn are seen. It is where we "put our best foot forward" and think of how we appear.

Nor is this awareness confined to the individual. As skill in walking—and other forms of pedestrian activity—becomes standardized and subject to aesthetic appraisal, a number of group disciplines evolve: dancing, racing, athletic competitions, processions, and parades become impressive displays of collective discipline and are viewed with pride by the community. So a wider definition of odology as (among other things) the study of our reaction to motion along a prescribed path can help us understand the social and aesthetic development of early man. It can also reintroduce the human and emotional response to every journey we take, even if it is on a superhighway.

One good reason for our having neglected the psychological and cultural roles of prehistoric paths is that such paths are hard to locate. Given a century or two, or even a year, they have a tendency to vanish; and in some parts of the world constant plowing and building and paving have either covered them up or transformed them beyond all recognition.

Moreover, it is clear that whatever paths might have existed were made obsolete many centuries ago by roads wide and smooth enough for wagons. The wheel became general some five thousand years ago in what is now southern Iraq. The wagon and chariot were soon invented and, when harnessed to the domesticated horse or ox, changed the way man traveled and lived. Wagons carrying produce from the countryside fostered the growth of towns and cities, and opened remote regions to commercial exploitation. The horse-drawn chariot made the conquest of foreign countries an easier undertaking. Rural paths serving small and often primitive communities ceased to play a historic role. Their obsolescence marked the beginning of modern Western history based on urbanization, imperialist expansion and continent-wide trade, and the spread of agriculture.

Historians and students of pretechnological Europe largely ignored the fact that the pedestrian way of life had lasted hundreds of thousands of years and had formed our ideas of community, of time and space and our relationship with the environment; that even after the advent of road systems and centralized governments it continued until well into modern times in many Old World societies. The evolution of the Old World landscape and of classical civilization can be understood only by giving priority to the vehicular road and the centralized state: what went before is of no more than archaeological interest.

It so happens, however, that the Western Hemisphere, from the time (some thirty thousand years ago) when the first settlers arrived from Asia until the early sixteenth century, produced a number of imposing and long-lasting cultures entirely without wheeled vehicles or domesticated beasts of burden and traction. When it at last produced engineered roads superior in construction to the roads of imperial Rome, those roads (entirely for pedestrian use) were essentially the ultimate stage in the development of the path. Thus, if we want to trace the evolution of the prehistoric path and its impact on culture, the place to start is the New World. We are fortunate in having retained a great many areas, in both South and Central America, where the pedestrian way of life persists, and even here in the United States the memory of the path is still alive, especially among the Indian communities of our Southwest.

We are learning to visualize pre-Columbian North America in a new and perhaps more accurate way than earlier generations did. Nineteenth-century writers described the area between the Atlantic and the Mississippi as a vast and impenetrable forest, inhabited by a scattering of nomadic Indians. The recent commemoration of the arrival of Columbus in the New World five centuries ago has prompted many historians and geographers to reexamine the evidence of what our eastern and southern and midwestern landscapes were like. The picture now emerging of what anthropologists call woodland and prairie America is of a landscape with extensive areas of farming: a landscape predominantly woodland but containing regions along the coast from Maine to North Carolina of villages with houses and gardens and cultivated fields in the midst of a parklike forest, with a network of well-traveled paths and trails. It was a landscape resembling in some ways that of northwestern Europe in the early seventeenth century, before roads and wheeled vehicles had become common.

Jacques Cartier in 1535 visited an Indian village near the present site of Montreal. He wrote that "as we went we found the way well beaten and frequented as can be, the fairest and best country that can possibly be seen, full of as goodly oaks as are in any wood in France. . . . About a mile and a half further we began to find goodly and large fields, full of such corn as the country yieldeth."[9] The Indian village was so large and impressive that Cartier called it a city.

Other travelers in the early seventeenth century reported much the same kind of landscape in Massachusetts and Virginia, and even in Florida, and for a brief moment it seemed that the Old World and the New shared an ancient tradition, one aspect of which was a pedestrian way of life. In less than a century, however, the illusion had totally vanished: the Indian farm landscape had been depopulated, and the contemporaneous rural landscape of Europe was being transformed by roads and a growing traffic of carriages and wagons. When woodland eastern America was invaded by colonists from England and France, the paths created by the Indian farmers and hunters were taken over by the newcomers and used as they had used the paths and lanes of the homeland. Finally the paths became crude roads for carts and wagons and horsemen, less and less traveled by Indians. But they survived, and eventually acquired a certain picturesque value. As a young man Hawthorne wrote a nostalgic essay on the disappearance of the Indian trails. "The forest track trodden by the hobnailed shoes of those sturdy Englishmen," he lamented, "has a distinctness which it never could have acquired from the light tread of a hundred times as many moccasins."[10] Cooper's novel *The Pathfinder* celebrated a different kind of path, and local historians throughout the East sought to locate and map what remained of a network which a nineteenth-century historian, with some exaggeration, described as extending from the southern point of Patagonia to the country of the Eskimos.

One of the most prolific students of the Indian path was the geographer Archer Butler Hulbert. Until his death in 1920, he continued to produce books and articles on the historic roads and highways in the eastern United States, and in 1902 he published a two-volume *History of Indian Thoroughfares.* His studies were confined to the paths in woodland America and he was primarily interested in the topographic and climatic factors determining their location; but in the course of his researches he accumulated much obscure information.

Hulbert's basic thesis was that many of the more heavily traveled paths used by the Indians were originally made by buffalo. The popular imagination still associates buffalo with the Great Plains west of the

Mississippi, but in fact they once ranged all over what is now the United States, and into Mexico and Canada. Well into the eighteenth century they were plentiful in the Midwest, especially in Kentucky and Tennessee and Ohio, and the first recorded description of a buffalo in the colonies dates from seventeenth-century Virginia. The wide trails produced by buffalo moving sometimes four abreast served to open the Midwest to white settlers through the Cumberland Gap and across New York State, and were known as "bison roads."

Although for their daily comings and goings the Woodland Indians used the narrower and shorter paths they had created, for their long-distance and interregional journeys they depended on the buffalo trails. Pre-Columbian Indians were, generally speaking, very mobile, constantly on the move as traders or hunters or migrants, or as sightseers. Always on foot, never pressed for time, and little burdened by possessions, some eastern tribes traveled to the Black Hills of South Dakota, and in the West Indians went hundreds of miles to get blankets from the Pueblos. A French traveler in Guiana in the eighteenth century noted with surprise that the local inhabitants went a hundred or two hundred leagues, "just to bring back a hammock or to attend a dance." An Englishman, captured by a band of southern Indians and obliged to stay with them for several months, told of how they were easily diverted from their planned itinerary: they made a day's detour to visit a friendly tribe, then on a sudden impulse went to raid a defenseless village. The possibility of trading with another village two hundred miles off their course, and later the chance to witness an Indian ceremony they had heard about but never seen, altogether prolonged their journey by another month.[11] So many pre-Columbian Indians enjoyed traveling—even on foot and over unfamiliar and sometimes primitive paths.

Hulbert sought to classify those paths by their function. He enumerated the local or village path, the path leading to the village farmlands, the path leading to the outside world—perhaps a buffalo trail—and finally the warpath. Except in the case of the last, all of these paths were similar in their total lack of any engineering or improvement. They wound through the forest or the grassland to avoid obstacles and uncertain terrain. They crossed streams at their narrowest width regardless of the detour. When a path was obstructed by a rock slide or a fallen tree, it went around it, and no one undertook to clear it. In winter, a path followed the ridge of hills where there was less brush and where the snow had been blown away. No sign indicated that a path was closed or abandoned. Despite a popular legend, no Indian path was ever marked for the benefit of travelers. The practice of blazing trees was confined to white pioneers. Indians either knew which direction to go or were capable of finding their way by themselves.

The only path modified to suit its traffic was the warpath. We use the term as a figure of speech, but in the pre-Columbian woodlands there were in all no more than four true warpaths leading from the territory of one tribe or confederation to the territory of an inveterate enemy; as permanent paths designed for that sole purpose, they were universally feared and avoided. Hulbert describes a typical warpath as "a deeper, wider, harder trail than any other early Indian thoroughfare, flanked by a thousand secret hiding places and lined with a long succession of open spots where warring parties were wont to camp." As soon as a tribe or community had resolved on war, the path became the scene of numerous rites and dances, and the participants underwent a change in status, becoming warriors, subject to a new kind of discipline and magic. An early explorer in Virginia wrote that "when they goe to warres they carry about with them their idol, of whom they ask counsel as the Romans were wont to do of the Oracle Apollo. They sing songs as they march towardes the battell in stead of drums and trumpets; their warres are very cruell and bloody."[12]

The dimensions and layout of the warpaths strongly suggest that Indian societies knew how to plan and produce artificial roads and spaces but did so only for significant collective religious or ceremonial events. For the carefully surveyed and constructed highway with uniform width and surface and a straight course regardless of topography may well have originated not from commercial pressure but from religious rites. The warpath, leading as it did to sacrifice and death, could possibly have been the earliest North American example of the formal, planned road common in other regions of pre-Columbian America.

Hulbert says nothing of any path with a religious or sacred association, yet it would be hard to find a primitive society, either in the New World or in the Old, which failed to celebrate some particular path as the one used by remote ancestors in the course of their legendary travels. Half a century ago the Blackfoot Indians still pointed out the trail from Canada down along the Rockies that their forebears had first traveled, and that they themselves used. As for the Pueblo Indians, their mythical history consists of little more than the chronicle of incessant movement from one spot, one village, to another. For their descendants to retrace that journey, to travel the same path, is seen as an act of piety, a way of returning to sacred origins. Throughout the primitive pedestrian world, the path has its beginnings when some hero ventures into the brush or forest where no one has previously been. He becomes a public benefactor, and as others follow in his footsteps, the path opens up a new world: it leads to the first village, to the first hunt; to ancestral shrines and to lands of promise; to war and new visions of nature. The path is so much a part of existence that it eventually becomes a metaphor for human life itself. Life is a road, long and unpredictable, and full of danger, that each of us must travel.

To the Navajo, as to many other Indian tribes, the metaphor is almost an article of faith. Spiritual preparation for adult existence emphasizes the journey ahead. "Man's life cycle," Gladys Reichard wrote, "is called a 'walk' through time. He travels a 'trail' repeatedly symbolized in sand painting and ritual. A major purpose of the ritual is to carry him safely and pleasantly along this road from birth to dissolution."[13] It is worth noting that the road itself is not subject to change or improvement; the emphasis is always on providing the traveler with guidance and protection: lines of corn pollen define the trail ahead and insure the traveler's safety.

In terms of the metaphor, "on the road of life to his final destiny which will make him one with the universe, Man is concerned with maintaining harmony with all things, with subsistence and the orderly replenishment of his own kind. . . . By himself man cannot annihilate error. Consequently, all beings in the sky, on the earth, in the waters under the earth . . . either aid him or must be overcome by superior power. Man may hope to obtain such power by proper manipulation, that is, by magical techniques."[14]

The Navajo journey is not, strictly speaking, a pilgrimage resulting in a transcendent experience; it is everyday existence ruled and protected by rites and precautions. He follows the paths his forebears followed. They lead to places where there are rare and useful herbs. They avoid all places associated with death. They lead to altars and shrines, to places of natural beauty and good hunting. The Navajo returns from the journey—whether it occupies a day or a lifetime—with food and resources for the family. He has paid tribute to the graves of ancestors, obeyed all injunctions and performed all necessary rites; above all, he has become familiar with his native landscape and become a harmonious element in it. The journey has made him a wiser and more knowledgable man; he has learned how to accept the path with all its hazards.

The interpretation of the road as a metaphor for our career here on earth has had many variations. The scientifically minded odologist will account for the European version by pointing out, quite correctly, that the nature and function of the path (or lane) had in the seventeenth century undergone a radical change. The prosperous element in European society—the merchants, the officials, the nobility, and the well-to-do landowners—had become accustomed to travel either on horseback or in coaches and carriages. The traditional rural path was often roughly and inefficiently enlarged and transformed into a primitive road for vehicles, and those traveling on foot were increasingly viewed as members of the lower class: they were footmen, footboys, footpads, foot soldiers; persons belonging at the foot of the social order; in critical writing, pedestrian came to mean labored, commonplace, and without style.

The European metaphorical use of the words *road* or *way* or *path* emphasized the difficulties encountered by the average wayfarer in the course of his or her journey through life. The most celebrated use in that new sense was in Bunyan's *Pilgrim's Progress.* Christian, the protagonist, sets out on a special kind of journey, not in order to satisfy daily needs or to conform to local tradition, but in order to reach a distant, highly desirable goal: salvation. As a metaphor of man's struggle to achieve redemption, the pilgrim's progress is a comprehensive analysis of Protestant theology: as an account of a lengthy journey through what was in effect seventeenth-century England, it vividly describes the obstacles, legal as well as topographical, which made every pedestrian journey a wearisome undertaking.

As long as those conditions persisted, as long as the average man or woman had to confront the indignities and complications of traveling on foot, the metaphorical message of Bunyan's work remained most important. But over the last century and a half, two developments have taken place: we have produced a new kind of road and a new metaphor, a vast network of smooth, efficient highways leading to every conceivable destination. At the same time we have largely ceased to believe in one universally accepted religious goal, usually identified with Christianity and the notion of spiritual redemption and of an afterlife. Heaven is no longer our destination. A third interpretation is taking shape: a multitude of roads, each with its own destination, obliges us to choose, to make decisions of our own; and the discourse of planning, of policy in the public realm, increasingly resorts to such road-associated phrases as *crossroads, dead ends, avenues of agreement, gridlock, collision course, impasse,* and *bypass.*

"Two roads diverged in a yellow wood / And sorry I could not travel both / And be one traveler . . . ." Robert Frost's poem "The Road Not Taken" has implications transcending the individual experience. It tells us of the dilemma of living in a world where there is no longer the one right way, the royal road to happiness and success, a path to the Heavenly City. Whichever road we take ultimately leads us to the agonizing moment of private decision. As with Saul of Tarsus, the road to Damascus may lie straight ahead, but it is only in the course of the journey that we discover our true destination.

(Author's drawing)

NOTES

Preface

1 From B. Mitchell, ed., *The Battle of Maldon and Other Old English Poems* (London, 1965); quoted in David M. Wilson, ed., *The Archeology of Anglo-Saxon England* (London: Methuen, 1976).

Chapter 1

1 A. Van Gennep, *Rights of Passage,* trans. Monika B. Vizedom and Gabrielle L. Cafee (Chicago: University of Chicago Press, 1960), 21.

Chapter 3

1 Cosmos Mindeleff, "Cliff Ruins of the Canyon de Chelly, Arizona," *Sixteenth Annual Report of the Bureau of American Ethnology* (Washington, D.C.: Smithsonian Institution, 1894–95), 90.

2 Benjamin Lee Whorf, "Linguistic Factors in the Terminology of Hopi Architecture," in John Carroll, ed., *Language, Thought, and Reality* (Cambridge: MIT Press, 1956), 200, 205.

3 C. Mindeleff, "Cliff Ruins," 59.

4 Ibid., 153.

Chapter 4

1 Whorf, "Linguistic Factors."

2 Victor Mindeleff, "A Study of Pueblo Architecture in Tusayau and Cibola," *Eighth Annual Report of the Bureau of American Ethnology* (Washington, D.C.: Smithsonian Institution, 1886–87), 140.

Chapter 6

1 Russell Lord, *The Care of the Earth* (New York: New American Library, 1962), 58.

2 William L. Thomas, Jr., et al., eds., *Man's Role in Changing the Face of the Earth* (Chicago: University of Chicago Press, 1960), 49.
3 Henry William Herbert, *Frank Forester's Field Sports of the United States, and the British Provinces of North America,* 2 vols. (New York, 1858).
4 Ibid.
5 Simon Schama, "Homelands," in *Home: A Place in the World,* special issue of *Social Research,* Spring 1991: 11.
6 Goethe, "Die Vereinigten Staaten," trans. Stephen Spender in Thomas Mann, ed., *The Permanent Goethe* (New York: Dial, 1948).
7 Richard Bridgman, *Dark Thoreau* (Lincoln: University of Nebraska Press, 1982).
8 See Barrington Moore, "Importance of Natural Conditions in National Parks," in G. B. Grinnell and Charles Sheldon, eds., *Hunting and Conservation: The Book of the Boone and Crockett Club* (New Haven: Yale University Press, 1925), 345–54.
9 C. S. Lewis, "The Four Loves," in *The Inspirational Writings of C. S. Lewis* (New York, 1987), 223.

Chapter 7

1 John Evelyn, *Sylva: or a Discourse of Forest Trees* (1664; London, 1679), 118.
2 Ibid., 114, 20.
3 Daniel Faucher, *Géographie agraire: Types des cultures* (Paris: Librairie des Médicis, 1949).
4 Evelyn, *Sylva.*

Chapter 8

1 Peter Collins, *Changing Ideals in Modern Architecture* (London: Faber and Faber, 1965), 50.
2 Charles Dickens, *Sketches by Boz* (1836; Everyman ed., 1907), 79ff.
3 Frank Cilla, ed., *Public Architecture in the United States* (New York, 1989), 12.

Chapter 9

1 John B. Jackson, *The Necessity for Ruins, and Other Topics* (Amherst: University of Massachusetts Press, 1980).
2 Percy Wells Bidwell and James I. Falconer, *History of Agriculture in the Northern U.S., 1620–1860* (Washington, D.C.: Carnegie Institution of Washington, 1925), 163.
3 Samuel Deane, *The New England Farmer: or Georgical Dictionary,* 3d ed. (Boston, 1822), 165.
4 *Rural Affairs* 4 (1878).
5 H. A. Kellar, *Solon Robinson, Pioneer and Agriculturist* (Indianapolis: Indiana Historical Bureau, 1936).

Chapter 10

1 Arien Mack, in *Home: A Place in the World,* special issue of *Social Research,* Spring 1991: 5.
2 Ibid., 135 (George Cateb) and 46 (John Hollander).
3 Ibid., 41 (Hollander), 81 (Breytenbach), 130 (Kim Hopper).
4 Yi Fu Tuan, *Segmented Worlds and Self* (Minneapolis: University of Minnesota Press, 1982), 68.
5 Ibid., 139, 181.
6 *Home: A Place in the World,* 274.
7 Carl Bridenbaugh, *The Colonial Craftsman* (New York: New York University Press, 1950), 39.
8 Thomas Hubka, *Big House, Little House, Back House, Barn* (Hanover, N.H.: University Press of New England, 1984); Yi Fu Tuan, *Segmented Worlds;* Philippe Ariès et al., eds., *A History of Private Life,* vol. 1 (Cambridge: Harvard University Press, 1987).
9 J. B. Jackson, "The Westward Moving House," in *Landscape: Selected Writings of J. B. Jackson* (Amherst: University of Massachusetts Press, 1970), 42.

Chapter 11

1 Eviatar Zerubavel, *Hidden Rhythms: Schedules and Calendars in Social Life* (Chicago: University of Chicago Press, 1981), xii.
2 Ibid., 141.
3 Paul Tillich, *Theology of Culture* (Oxford: Oxford University Press, 1959), 16.

Chapter 13

1 John Bell Rae, *The Road and the Car in American Life* (Cambridge: MIT Press, 1971), 255.

Chapter 14

1 Joseph Rykwert, *Adam's House in Paradise* (New York: Museum of Modern Art, 1972).
2 Edgar Anderson, *Plants, Man, and Life* (Boston: Little, Brown, 1967), 9.
3 Ibid., 136–50.
4 Ibid.
5 William L. Thomas, Jr., et al., eds., *Man's Role in Changing the Face of the Earth* (Chicago: University of Chicago Press, 1960), 766.
6 Ibid., 767.
7 Anderson, *Plants, Man, and Life,* 146.
8 Thomas, *Man's Role,* 777.
9 David B. Quinn, ed., *North American Discovery, Circa 1000–1612* (Charleston: University of South Carolina Press, 1971), 103ff.
10 Nathaniel Hawthorne, *Main Street* (Boston: Houghton Mifflin, 1883).
11 William E. Myer, "Indian Trails of the Southeast," *Forty-second Annual Re-*

port of the Bureau of American Ethnology (Washington, D.C.: Smithsonian Institution, 1924–25), 727–857.

12 Archer Butler Hulbert, *History of Indian Thoroughfares* (Cleveland: A. H. Clark Co., 1902), 49.

13 Gladys A. Reichard, *Navaho Religion: A Study of Symbolism* (Princeton: Princeton University Press, 1974), 37.

14 Ibid., 49.

INDEX